MAGNETIC FRINGE FIELDS AND INTERFERENCE IN HIGH INTENSITY ACCELERATORS

To Martin:

Best wishes for good health
and
happy retirement life.

— J. G.
September 1, 2010
Oak Ridge, TN

MAGNETIC FRINGE FIELDS AND INTERFERENCE IN HIGH INTENSITY ACCELERATORS

J. G. WANG

Nova Science Publishers, Inc.
New York

For permission to use material from this book please contact us:
Telephone 631-231-7269; Fax 631-231-8175
Web Site: http://www.novapublishers.com

LIBRARY OF CONGRESS CATALOGING-IN-PUBLICATION DATA

Wang, J. G. (Jian Guang)
Magnetic fringe field and interference in high intensity accelerators / J.G. Wang.
p. cm.
Includes bibliographical references and index.
ISBN 978-1-60876-946-9 (softcover)
1. Neutrons--Scattering. 2. Spallation (Nuclear physics) 3. Proton accelerators. I. Title.
QC793.5.N4628W36 2009
539.7'3--dc22
2009049848

Published by Nova Science Publishers, Inc. † New York

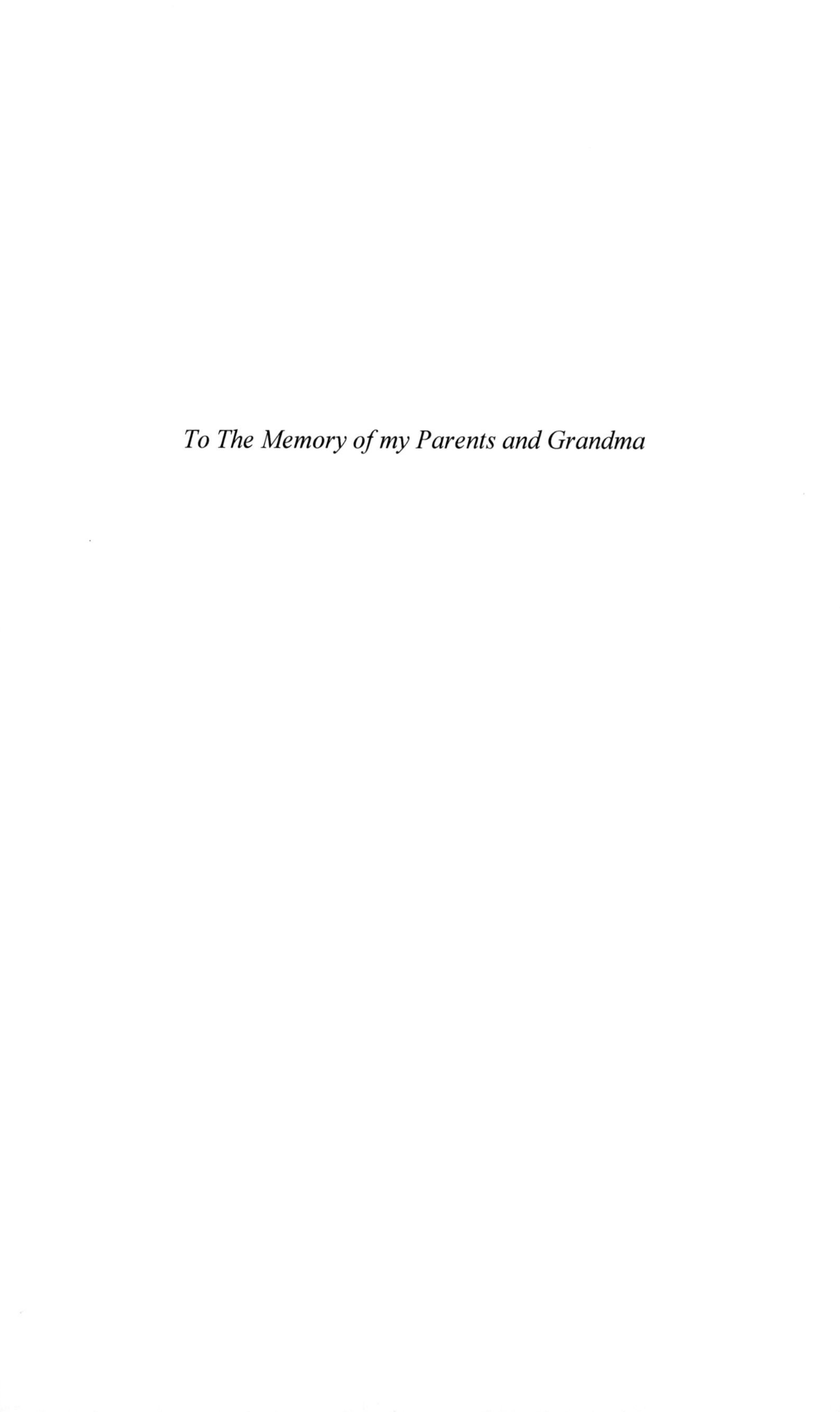

To The Memory of my Parents and Grandma

CONTENTS

PREFACE

This monograph has evolved from my research work on the Spallation Neutron Source (SNS) at Oak Ridge National Laboratory (ORNL). The SNS is the world's most powerful accelerator-driven, short-pulse neutron scattering facility for scientific research and industrial development. It consists of a powerful proton accelerator, a liquid mercury target, and neutron scattering instruments. The SNS currently holds the world record for proton beam power for pulsed spallation sources, 1 megawatt, and is expected to achieve the design power of 1.4 megawatts by the summer of 2010. The SNS accelerator complex consists of a 2.5 MeV H^- beam injector that includes a high-brightness H^- ion source and a radio frequency quadrupole, a high-power linear accelerator (linac) of up to 1 GeV energy with a pulse length of 1 ms at 60 Hz, and an accumulator ring to compress the long linac pulses into short ones of about 700 ns with 1.5E14 protons per pulse.

The design and construction of the SNS project was a partnership of six national laboratories of the U.S. Department of Energy (DOE). They were Lawrence Berkeley National Laboratory, responsible for the design and development of the H^- ion source and injector; Los Alamos National Laboratory and Jefferson Laboratory for the SNS linac, including a room temperature linac (drift tube linac and coupled cavity linac) followed by a superconducting linac (SCL); Brookhaven National Laboratory (BNL) for the accumulator ring plus two beam transport lines; ORNL for the project site for the mercury target system and all conventional facilities; and Argonne National Laboratory and ORNL for neutron scattering instruments.

I joined ORNL in 1998 after several years of research work for Prof. M. Reiser at the University of Maryland–College Park, where we studied the physics of space-charge dominated beams and produced very fruitful results. The original plan for me at ORNL was to participate in the design and development of the SNS accumulator ring and to prepare for its commissioning and operation. Thus I was stationed at BNL to work with the BNL/SNS team. At BNL, I familiarized myself with the design of the SNS accumulator ring. In particular, I studied beam instabilities in the ring to be built and measured both longitudinal and transverse coupling impedances of a few SNS ring components, such as a transmission line beam position monitor and a fast extraction kicker magnet. The measurement results led to a later redesign of the SNS ring extraction kickers.

After about a year at BNL, I moved back to ORNL to perform SNS magnet measurements on site. We started from scratch to establish a magnet measurement lab at ORNL and completed about 400 magnets for the SNS linac and the ring transport lines during the SNS construction period. At the same time I learned to use OPERA/TOSCA code for magnet modeling and analyses. Gradually, I realized that there were still important issues, such as magnetic fringe fields and interference, in the design of large aperture magnets for high-current accelerators. Since the SNS commissioning and operation in 2006, I have been involved mainly in solving beam dynamics problems in the SNS ring injection and extraction regions. These include huge particle losses in the ring injection waste beam dump line, beam profile distortion on the target due to an extraction septum magnet, particle trajectories in the injection momentum collimation dump line, and beam quality degradation and particle losses in the SCL. A key to all these problems is how to correctly and accurately handle the magnets involved.

Magnetic devices such as solenoids, dipoles, quadrupoles, sextupoles, wigglers, and undulators are essential components for beam orbit control in charged particle accelerators. Magnetic fringe fields and their interference play important roles in particle dynamics, but their analysis is not always straightforward. In high-energy accelerators in which the magnets usually have very large aspect ratios (length over aperture), the contribution of magnetic fringe fields to particle motion is relatively or even negligibly small; therefore, simple approximations usually suffice. Since most of the magnets are installed far apart, magnetic interference rarely occurs. However, this is not the case for advanced high-intensity accelerators, especially for high-current circular machines such as those for light sources and spallation neutron sources. Magnetic devices in these machines usually have a large aperture and short length (small aspect ratio), and they are packed densely along beam lines as a result of limited space. The fringe fields of magnets become more profound, and the interference among adjacent magnets often causes problems for machine operation. This becomes a challenge for the design of high-intensity accelerators.

Magnetic fringe fields and their interference are usually difficult problems to analyze accurately because they have a highly nonlinear nature and generally require accurate 3-dimensional calculations. In recent years, because of progress in advanced magnet simulation codes and ever-increasing computing power, the analysis of magnetic fringe fields and interference can be handled with a reasonable effort to achieve high accuracy. In this book we will review the historical background and status of the treatment of magnetic fringe fields and interference in particle accelerators. Particularly, we will present recent research results on subjects associated with the SNS accumulator ring. Magnetic field distributions of the SNS ring magnets are analyzed at high accuracy for their fringe constituents based on 3-dimensional computer simulations. Taking into account the effects of magnetic fringe fields and interference, we obtain more accurate particle optics in the magnets; our results differ considerably from conventional treatments. Particle tracking directly in simulation models containing magnet assemblies provides a new and better approach to studying beam dynamics in beam lines. The methods and techniques developed in our research should be directly applicable to other high-intensity accelerators.

The preparation of this book would not have been possible without the support and encouragement of J. Haines, G. Murdoch, and D. Lousteau. I am deeply grateful for their advice and support of my research work over so many years. I am also indebted to D. Lousteau for his review of the entire book manuscript. I would like to express my gratitude to the following people at the SNS for their support and assistance during the work which led to the material presented in this book: J. Error, J. Galambos, S. Henderson, J. Holmes, T. Hunter, M. Plum, T. Roseberry, and Y. Zhang. I have received many valuable ideas and suggestions from stimulating discussions with BNL colleagues A. Jain, J. Jackson, W. Z. Meng, and N. Tsoupas. I am greatly indebted to many people in Vector Fields Software, particularly to J. Simkin and C. Riley, for their help in OPERA/TOSCA applications. I would like to take this opportunity to express my appreciation to Nova Science Publishers, Inc., and its editors whose work has made the publication of this book possible. Finally and most important, I want to thank my wife Mei-li for her enthusiastic support during my research career and my son Andrew for his proofreading of this book.

J. G. (Jian-Guang) Wang
September 2009
Oak Ridge, Tennessee, USA

Chapter 1

INTRODUCTION

1.1. BACKGROUND

Magnetic devices such as solenoids, dipoles, quadrupoles, sextupoles, wigglers, and undulators are essential components for beam orbit control in charged particle accelerators. Magnetic force is long range in nature, and magnetic fringe fields from these devices can extend for a long way and interfere with adjacent magnets and other magnetic materials. Magnetic fringe fields and their interference play important roles in particle dynamics, but their analysis has been a challenging task. The study of magnetic fringe fields in charged particle beam lines and accelerators has a history as long as that of the magnetic devices themselves. Early references to magnetic optics and magnetic fringe fields can be found in monographs [1–4]. Although magnetic fringe fields in accelerator magnets have been well investigated [5–16], it appears that their effects on beam dynamics and machine operation are often overlooked in design practice. This lack of attention may not be unreasonable for high-energy accelerators, but that is not the case for high-intensity machines. Studies of the topic of magnetic interference among accelerator magnets are so far rarely seen in the literature.

In high-energy accelerators, which employ mostly superconducting magnets with very large aspect ratios (length over aperture), the contribution of magnetic fringe fields to particle motion is relatively or even negligibly small. Simple approximations in magnet optics often suffice. In this scenario, conventional hard edge models in two dimensions, which do not take into account magnetic fringe fields for lattice elements, are widely used. Since most of the magnets in high-energy accelerators are also installed far apart, magnetic interference rarely exists. Only in some specific locations where magnets are more densely packed—such as beam injection, extraction, and collision—must special care be taken to account for magnetic fringe field and interference problems.

In contrast, advanced high-intensity accelerators, especially high-current circular machines such as those used in light sources and spallation neutron sources, face additional challenges related to magnet optics. High-intensity accelerators are low-energy or medium-energy particle accelerators and employ conventional iron-dominated electromagnets. Magnetic devices in these machines usually are short in length with large apertures (small aspect ratios), and they are packed densely along beam lines as a result of limited space. In such a scenario, the fringe fields of magnets become more profound, and interference among adjacent magnets often causes problems in machine commissioning and operation. It was reported during the inauguration of the COSY cooler synchrotron at Juelich, Germany, that

> Although all magnetic components had been precisely measured in bench tests, it was apparent from the beginning that the close mounting inside the ring would affect their magnetic properties. This, and the large number of elements involved, made the commissioning process tough [17].

How to accurately and effectively handle the problems of magnetic fringe fields and interference in high-intensity accelerators remains in question.

1.2. Research Issues

The accumulator ring of the Spallation Neutron Source (SNS) at Oak Ridge National Laboratory is a typical example illustrating the effects of magnetic fringe fields and interference in high-intensity machines. The SNS accelerator is designed to achieve an unprecedented beam intensity of 1.5E14 protons per pulse in its accumulator ring, but with a relatively low energy of 1 GeV [18–20]. The magnet aperture has to be made very large to provide a betatron acceptance of 480 π mm mrad, which is three times the un-normalized beam emittance. Each magnet is short with an aspect ratio as small as 1.5. Magnetic fringe fields are profound in all the SNS ring magnets. A total of close to 200 magnetic dipoles, quadrupoles, sextupoles, and correctors are densely packed along a ring with a circumference of merely 248 meters. This creates significant interference among the devices [21]. Unfortunately, the problems of magnetic fringe fields and interference in the SNS ring magnets were not adequately studied in the design stage; therefore, they have caused much difficulty in machine commissioning and operation.

In the SNS ring injection region, four chicane dipoles and an injection dump septum magnet are installed in close proximity, and their fields overlap. The injection design was based on isolated, simplified 2-dimensional hard edge models for its lattice components [22, 23]. The problem was in fact a true 3-dimensional one. During the first year of SNS ring operation, it was not possible to simultaneously transport both unstripped and partially stripped H^- beams through the waste beam dump line. Activation levels in this region were the highest in the SNS accelerator complex [24–26]. Several design issues led to the difficulties during the SNS ring commissioning and early operation. Magnetic fringe fields and associated interference in the injection region were a key factor leading to heavy particle losses in the injection dump beam line [27, 28]. It became a key issue to correctly simulate the magnet assembly and accurately compute particle trajectories of the waste beams.

During the SNS ring commissioning, the beam profile on the target was found to be slightly tilted. This mismatch between the extracted beam profile and the target cross section had been an issue of concern for high-power operation [24, 26]. A considerable amount of effort had been devoted to finding the source of and the remedy to the problem. Simulation studies showed that a Lambertson septum magnet in the SNS ring extraction region contained a rather large skew quadrupole component, which was dominated by the contribution from the fringe fields at its entrance and exit [29, 30]. This large skew quad term resulted in the beam profile distortion of the extracted beam. The effect was further enhanced by the interference from an adjacent magnetic quadrupole.

The SNS ring lattice design was based on conventional hard edge models for its components, as is usually the case in high-energy accelerators. This practice is not sufficiently accurate because the SNS lattice magnets have small aspect ratios and are densely packed. Their correct optics differ from those predicted by conventional treatments based on simplified hard edge models [31]. There have been observations of the discrepancies between the lattice design predictions and the SNS ring operation performance in terms of the linear optics such as the betatron tune [32]. Finding the real sources of the discrepancies is a topic of further investigation.

Even in the SNS linac, magnetic fringe fields leave their footprint in deteriorating beam quality and particle losses. The SNS linac system is designed to deliver 1 GeV pulsed H^- beams up to an unprecedented power of 1.56 MW. A major design criterion is to keep the relative fractional particle losses below the 1×10^{-4} level in order to allow hands-on maintenance. The magnets for the SNS linac lattice have small aspect ratios and extended fringe fields. Thus they generate nonlinear magnetic fields, such as the pseudo-octupole term, leading to the third-order aberration. In addition, the SNS linac magnets contain a rather large dodecapole term, mainly produced by the fringe fields at magnet ends. The dodecapole term leads to the fifth-order aberration. Although the effect on the beam dynamics from the higher-order aberrations of each magnet is small, its accumulation over a large quantity of magnets is detrimental to the beam. Preliminary studies have shown that, with the nonlinear dodecapole term, the beam transverse distribution in the linac is degraded in developing longer Gaussian tails, which would cause particle losses in the injection region. Moreover, these nonlinear fields have already caused particle losses at a 10^{-4} level in the SNS superconducting linac [33, 34].

Magnetic fringe fields and their interference are usually difficult problems to analyze accurately because the problems typically are highly nonlinear and involve 3-dimensional calculations or measurements. In recent years, with progress in advanced magnet simulation codes and ever-increasing computing power, it has been possible to compute magnetic fringe fields and interference at high accuracy. In this book, we briefly review the historical background and status of the treatment of magnetic fringe fields and interference in particle accelerators. In particular, the recent research results on the subject associated with the SNS accumulator ring are presented. Magnetic field distributions of the SNS ring magnets are analyzed from 3-dimensional computer simulations at high accuracy for their fringe constituents. Taking into account the effects of magnetic fringe fields and interference, more accurate particle optics for the magnets are obtained, which differ considerably from conventional treatments. We also perform 3-dimensional particle tracking directly in simulation models involving magnet assemblies. This is an even better approach to studying beam dynamics in short beam lines. The methods and techniques developed in the research should be directly applicable to other high-intensity accelerators.

1.3. Synopsis

The material presented in this book is mainly from several papers published in *Physical Review Special Topics–Accelerators and Beams* during the past several years. This book is organized as follows:

Chapter 2 provides the fundamentals of accelerator magnets for the convenience of readers to help them understand the research results described in subsequent chapters. The basics include static magnetic fields; 2-dimensional field presentations; particle motion in magnetostatic fields; the most widely used accelerator magnets, such as dipole magnets and quadrupole magnets; and magnet measurements. Chapter 3 briefly introduces computer codes and modern techniques of magnet modeling. A comprehensive set of references, though far from complete, is supplied for readers interested in studying the topic further. Specifically, an overview of the magnet design code OPERA-3D/TOSCA is presented. This tool is widely used in many national accelerator labs and other institutions. Finally, a modeling example with the SNS injection chicane dipoles is illustrated.

Chapter 4 is devoted to 3-dimensional multipole field expansion based on simulated field data. This is the foundation for deriving particle optics of magnets with realistic field distributions. The particle optics of a single quadrupole are analyzed in Chapter 5. The effect of magnetic fringe fields is fully included in the analyses, and the inaccuracy of conventional hard edge models is demonstrated.

In Chapter 6, we study magnetic interference between a quadrupole and a dipole corrector. The interference can be treated as a first-order perturbation to the original quad and can be represented by a small hard edge model. The particle optics in a quadrupole assembly consisting of a quadrupole doublet and a dipole corrector are given in Chapter 7. It is shown that the lens parameters derived from the rigorous first-order theory with realistic field distributions differ significantly from those due to conventional treatments.

In Chapter 8, we present 3-dimensional particle tracking in the SNS ring injection dump beam line, which suffered high particle losses in the SNS ring commissioning and early operation. Three-dimensional modeling of four closely packed magnets in the beam line and 3-dimensional particle tracking in the models has revealed the mechanisms of particle losses and provided remedies for the problem. In Chapter 9, a special problem with the SNS ring extraction Lambertson septum magnet is described. The original Lambertson septum contained a large skew quad term, which distorted the beam profile on the SNS target. The skew quad term is mainly produced in the fringe regions at the magnet's two ends. The detailed analyses and the solutions to the problem are presented.

REFERENCES

[1] K. G. Steffen, *High Energy Beam Optics*, John Wiley and Sons, Inc., New York, 1965.

[2] P. Grivet, *Electron Optics*, Pergamon Press Ltd, London, 1965.

[3] P. W. Hawkes, *Quadrupole Optics*, Springer-Verlag, Heidelberg, 1966.

[4] H. Wollnik, *Optics of Charged Particles*, Academic Press, New York, 1987.

[5] H. A. Enge, "Effect of extended fringe fields on ion-focusing properties of deflecting magnets," *Review of Scientific Instruments*, 35(1), 278 (1964).

[6] H. Wollnik and H. Ewald, "The influence of magnetic and electric fringe fields on the trajectories of charged particles," *Nucl. Instrum. Methods* 36 93 (1965).

[7] G. E. Lee-Whiting, "End effects in first-order theory of quadrupole lenses," *Nucl. Instrum. Methods* 76 305 (1969).

[8] G. E. Lee-Whiting, "Third-order aberrations of a magnetic quadrupole lens," *Nucl. Instrum. Methods* 83 232 (1970).

[9] H. Matsuda and H. Wollnik, "Third order transfer matrices of the fringe field of an inhomogeneous magnet," *Nucl. Instrum. Methods* 77 283 (1970).

[10] H. Matsuda and H. Wollnik, "Third order transfer matrices for the fringe field of magnetic and electrostatic quadrupole lenses," *Nucl. Instrum. Methods* 103 117 (1972).

[11] G. E. Lee-Whiting, "First- and second-order motion through the fringing field of a bending magnet," *Nucl. Instrum. Methods* A294 31 (1990).

[12] J. Irwin and Chun-Xi Wang, "Explicit soft fringe maps of a quadrupole," in Proceedings of the 1995 Particle Accelerator Conference, Dallas, TX, May 1–5, 1995, ed. L. Gennari (IEEE Piscataway, NJ, 1996), p. 2376.

[13] M. Venturini, "Scaling of third-order quadrupole aberrations with fringe field extension," in Proceedings of the 1999 Particle Accelerator Conference, New York, March 29–April 2, 1999, eds. A. Luccio and W. Mackay (IEEE Piscataway, NJ, 1999), p.1590.

[14] M. Berz, B. Erdelyi, and K. Makino, "Fringe field effects in small rings of large acceptance," *Phys. Rev. ST Accel. Beams* 3, 124001 (2000). (http://prst-ab.aps.org/abstract/PRSTAB/v3/i12/e124001)

[15] Y. Papaphilippou, J. Wei, and R. Talman, "Deflection in magnet fringe fields," *Phys. Rev. E* 67, 046502 (2003).

[16] R. Baartman and D. Kaltchev, "Short quadrupole parameterization," in Proceedings of the 2007 Particle Accelerator Conference, Albuquerque, NM, June 25–29, 2007 (IEEE Catalog Number: 07CH37866, 2007), p. 3229.

[17] "JUELICH COSY acceleration and cooling," *CERN Courier*, p. 3–4, October 1993.

[18] J. R. Alonso for the SNS team, "The Spallation Neutron Source project," in Proceedings of the 1999 Particle Accelerator Conference, New York, March 29–April 2, 1999, eds. A. Luccio and W. Mackay (IEEE Piscataway, NJ, 1999), p. 574.

[19] J. Wei, D. T. Abell, J. Beebe-Wang, M. Blaskiewicz, P. R. Cameron, N. Catalan-Lasheras, G. Danby, A. V. Fedotov, C. Gardner, J. Jackson, Y. Y. Lee, H. Ludewig, N. Malitsky, W. Meng, Y. Papaphilippou, D. Raparia, N. Tsoupas, W. T. Weng, R. L. Witkover, and S. Y. Zhang, "Low loss design for the high-intensity accumulator ring of the Spallation Neutron Source," *Phys. Rev. ST Accel. Beams* 3, 080101 (2000). (http://prst-ab.aps.org/abstract/PRSTAB/v3/i8/e080101)

[20] S. Henderson, "Status of the Spallation Neutron Source: Machine and Science," in Proceedings of the 2007 Particle Accelerator Conference, Albuquerque, NM, June 25–29, 2007 (IEEE Catalog Number: 07CH37866, 2007), p. 9.

[21] Y. Papaphilippou, Y. Y. Lee, and W. Meng, "Impact of magnetic field interference in the SNS ring," in Proceedings of the 2001 Particle Accelerator Conference, Chicago, IL, June 18–22, 2001, eds. P. Lucas and S. Webber (IEEE Piscataway, NJ, 2001), p. 1667.

[22] J. Wei, J. Beebe-Wang, M. Blaskiewicz, J. Brodowski, A. Fedotov, C. Gardner, Y.Y. Lee, D. Raparia, V. Danilov, J. Holmes, C. Prior, G. Rees, S. Machida, "Injection choice for Spallation Neutron Source ring," ibid, p. 2560.

[23] D. Raparia for the SNS collaboration, "SNS injection and extraction devices," in Proceedings of the 2005 Particle Accelerator Conference, Knoxville, TN, May 16–20, 2005, ed. C. Horak (IEEE Catalog Number 05CH37623C, 2005), p. 553.

[24] M. A. Plum, "Commissioning of the Spallation Neutron Source accelerator systems," in Proceedings of the 2007 Particle Accelerator Conference, Albuquerque, NM, June 25–29, 2007 (IEEE Catalog Number: 07CH37866, 2007), p. 2603.

[25] J. A. Holmes, M. A. Plum, J. G. Wang, Y. Zhang, and M. Perkett, "ORBIT injection dump simulations of the H^0 and H^- Beams", ibid, p. 3657.

[26] J. A. Holmes, S. Cousineau, M. P. Plum, and J. G. Wang, "Computational beam dynamics studies for improving the ring injection and extraction systems in SNS," in Proceedings of the Eleventh European Particle Accelerator Conference, Genoa, Italy, June 23–27, 2008, eds. I. Andrian et al. (EPS–AG, 2008), p. 3008.

[27] J. G. Wang, "3D modeling of SNS ring injection dump beam line", in Proceedings of the 2007 Particle Accelerator Conference, Albuquerque, NM, June 25–29, 2007 (IEEE Catalog Number: 07CH37866, 2007), p. 3660.

[28] J. G. Wang and M. Plum, "Three-dimensional particle trajectories and waste beam losses in injection dump beam line of Spallation Neutron Source accumulation ring," *Phys. Rev. ST Accel. Beams* 11, 014002 (2008). (http://prst-ab.aps.org/abstract/PRSTAB/v11/i1/e014002); also see J. G. Wang, SNS-NOTE-MAG-178, August 21, 2007.

[29] J. G. Wang, "Performance improvement of an extraction Lambertson septum magnet in the Spallation Neutron Source accumulator ring," *Phys. Rev. ST Accel. Beams* 12, 042402 (2009). (http://prst-ab.aps.org/abstract/PRSTAB/v12/i4/e042402); also see J. G. Wang, SNS-NOTE-MAG-183, March 13, 2009.

[30] J. G. Wang, "3D field quality studies of SNS ring extraction Lambertson septum magnet," in Proceedings of the 2009 Particle Accelerator Conference, Vancouver, B. C., Canada, May 4–8, 2009, ed. M. Comyn, MO6PFP031.

[31] J. G. Wang, "Particle optics of quadrupole doublet magnets in Spallation Neutron Source accumulator ring," *Phys. Rev. ST Accel. Beams* 9, 122401 (2006). (http://prst-ab.aps.org/abstract/PRSTAB/v9/i12/e122401); also see J. G. Wang, SNS-NOTE-MAG-168, July 12, 2006.

[32] Z. Z. Liu, S. M. Cousineau, J. A. Holmes, J. D. Galambos, and M. Plum, "Linear optics measurement and correction in the SNS accumulator," in Proceedings of the 2009 Particle Accelerator Conference, Vancouver, B. C., Canada, May 4–8, 2009, ed. M. Comyn, TH6PFP058.

[33] Y. Zhang, C. K. Allen, J. D. Galambos, J. Holmes, and J. G. Wang, "Beam transverse issues at the SNS linac," ibid, TH6PFP089.

[34] Y. Zhang, C. K. Allen, J. D. Galambos, J. Holmes, and J. G. Wang, "Transverse beam resonance in the superconducting linac of the Spallation Neutron Source," submitted to *Phys. Rev. ST Accel. Beams* for publication.

Chapter 2

FUNDAMENTALS OF ACCELERATOR MAGNETS

The SNS accelerator employs iron-dominated electromagnets rather than superconducting magnets. The most SNS magnets are driven by static electric currents. The problems with magnetic fringe fields and interference analyzed in this book are illustrated with examples based on the static SNS electromagnets. In this Chapter, we first provide the basic theory on static magnetic fields, which governs the performance of static electromagnets. Every textbook on electromagnetics should cover the topic. Only two references [1, 2] are listed. For many accelerator magnets, a two-dimensional representation of their fields is a good approximation. This is described in Section 2.2. Accelerator magnets are often treated as lenses for charged particles. The equations of motion of charged particles in static magnetic fields are essential for deriving particle optics. This is briefly introduced in Section 2.3. Much more information on beam dynamics can be found in many books such as in references [3, 4]. In Sections 2.4 and 2.5, we describe two most widely used accelerator magnets: dipoles and quadrupoles. A good reference for these and other iron-dominated electromagnets is [5]. In the last Section of this Chapter, magnetic measurements are presented based on our experience in the SNS magnets. All the material provided in this Chapter will serve as a prerequisite for subsequent chapters.

2.1. MAGNETOSTATIC FIELDS

Electromagnetism is governed by Maxwell's equations. For static magnetic fields in free space, Maxwell's equations reduce to specifying the divergence and the curl of the magnetic flux density vector, **B**, as

$$\nabla \bullet \mathbf{B} = 0 \tag{2.1a}$$

and

$$\nabla \times \mathbf{B} = \mu_0 \mathbf{J} . \tag{2.1b}$$

Here $\mu_0 = 4\pi \times 10^{-7}$ (H/m) is the permeability of free space and **J** (A/m^2) is the current density. In an integral form, Eq. (2.1a) becomes the law of conservation of magnetic flux

$$\oint_{\mathbf{s}} \mathbf{B} \bullet d\mathbf{s} = 0 , \tag{2.2a}$$

where the surface integral is carried out over the bounding surface **s** of an arbitrary volume. Similarly, Eq. (2.1b) leads to the Ampere's circuital law

$$\oint_C \mathbf{B} \bullet d\vec{\ell} = \mu_0 I , \tag{2.2b}$$

where the path C for the line integral is the contour bounding an open surface, and I is the total current through the surface.

The divergence-free of **B**, as depicted by Eq. (2.1a), assures that **B** is solenoidal and can be expressed as the curl of another vector field **A**, such that

$$\mathbf{B} = \nabla \times \mathbf{A} \text{ (T)}. \tag{2.3}$$

The vector field **A** is called the vector magnetic potential in Wb/m. With the choice of Coulomb gauge

$$\nabla \bullet \mathbf{A} = 0 , \tag{2.4}$$

we obtain a vector Poisson's equation

$$\nabla^2 \mathbf{A} = -\mu_0 \mathbf{J} . \tag{2.5}$$

Its solution is

$$\mathbf{A} = \frac{\mu_0}{4\pi} \int_{V'} \frac{\mathbf{J}}{r} dv' \text{ (Wb/m)}. \tag{2.6a}$$

For a direct current I flowing in a closed path C′, Eq. (2.6a) becomes

$$\mathbf{A} = \frac{\mu_0 I}{4\pi} \oint_{C'} \frac{d\vec{\ell}'}{r} \text{ (Wb/m).} \tag{2.6b}$$

By applying Eqs. (2.6a, b) to Eq. (2.3), one obtains

$$\mathbf{B} = \frac{\mu_0}{4\pi} \oint_{V'} \frac{\mathbf{J} \times \mathbf{r}}{r^3} dv' \text{ (T)} \tag{2.7a}$$

or

$$\mathbf{B} = \frac{\mu_0 I}{4\pi} \oint_{C'} \frac{d\vec{\ell}' \times \mathbf{r}}{r^3} \text{ (T).} \tag{2.7b}$$

Equation (2.7) is known as Biot-Savart law, which is often used to calculate steady magnetic fields from given direct current distributions.

In a current-free region $\mathbf{J} = 0$, the magnetic flux density $\mathbf{B}$ becomes curl-free, as indicated by Eq. (2.1b), and can be expressed as the gradient of a scalar field

$$\mathbf{B} = -\mu_0 \nabla \Phi_m, \tag{2.8}$$

where Φ_m is called the scalar magnetic potential. Since the magnetic flux density $\mathbf{B}$ is also divergence-free as indicated by Eq. (2.1a), one obtains

$$\nabla^2 \Phi_m = 0. \tag{2.9}$$

This equation is known as Laplace's equation.

In the presence of magnetic materials in an externally applied magnetic field, the materials are magnetized. The degree of magnetization is expressed by a magnetization vector $\mathbf{M}$, which is the volume density of magnetic dipole moment. A new fundamental field quantity, the magnetic field intensity $\mathbf{H}$, is then defined as

$$\mathbf{H} = \frac{\mathbf{B}}{\mu_0} - \mathbf{M} \text{ (A/m).} \tag{2.10}$$

The magnetic flux density **B** and the magnetic field intensity **H** are also related by

$$\mathbf{B} = \mu\mathbf{H} = \mu_r \mu_0 \mathbf{H}, \tag{2.11}$$

where μ_r is known as the relative permeability of the medium and μ is the absolute permeability.

In regions having media with different physical properties, **B** and **H** vectors must satisfy certain boundary conditions at the interfaces of different media. The first boundary condition is that the normal component of **B** is continuous across an interface. The second one is that the tangential component of **H** is continuous across the boundary of almost all physical media; it is discontinuous only when an interface with an ideal perfect conductor or superconductor is assumed. In magnet simulations, it is often to determine the boundary conditions by symmetry properties of magnet configurations. For instance, if a magnet has a mirror symmetry with respect to the x-y plane, the z-component of magnetic field must vanish on the plane and the boundary condition on the plane must be tangential.

2.2. Two-Dimensional Fields

For an accelerator magnet such as a dipole or a quadrupole, whose length is much longer than its aperture, the magnetic field in the central region is essentially two-dimensional: the transverse components of the field are uniform and the longitudinal component vanishes. Most magnets in high-energy accelerators tend to be very long compared to their aperture. A commonly used model for accelerator magnets is known as the hard edge model. The magnetic field along the hard edge length L is a constant and vanishes abruptly at magnet ends. This is also a two dimensional field and is a good approximation for many accelerator magnets. The transverse field components in a two-dimensional field satisfy the two-dimensional Laplace's equation (2.9). The components of **B** in cylindrical coordinates can be written in the form [6]:

$$B_r(r,\theta) = \sum_{m=1}^{\infty} C_m \left(\frac{r}{R}\right)^{m-1} Sin[m(\theta - \alpha_m)], \tag{2.12a}$$

$$B_\theta(r,\theta) = \sum_{m=1}^{\infty} C_m \left(\frac{r}{R}\right)^{m-1} Cos[m(\theta - \alpha_m)]. \tag{2.12b}$$

Here the integer m is the order of the multipoles. For example, m = 1 corresponds to a pure dipole, and m = 2 to a quadrupole, etc. C_m and α_m are the field amplitude and phase angle of the 2m-pole term of the total field, and R is a reference radius within the magnet aperture. Equations (2.12a, b) can also be written as

$$B_r(r,\theta) = \sum_{m=1}^{\infty} \left(\frac{r}{R}\right)^{m-1} \left[B_m Sin(m\theta) + A_m Cos(m\theta)\right], \quad (2.13a)$$

$$B_\theta(r,\theta) = \sum_{m=1}^{\infty} \left(\frac{r}{R}\right)^{m-1} \left[B_m Cos(m\theta) - A_m Sin(m\theta)\right], \quad (2.13b)$$

where

$$B_m = C_m Cos(m\alpha_m), \quad (2.14a)$$

$$A_m = -C_m Sin(m\alpha_m) \quad (2.14b)$$

are known as the 2m-pole normal and skew terms according to the "European Convention", which we follow in this book. An accelerator magnet, e. g. a magnetic dipole, quadrupole, etc., is called a normal magnet if $A_m = 0$, while it is called a skew magnet if $B_m = 0$. In a normal magnet, the magnetic fields are vertical along the horizontal axis. In a skew magnet, the magnetic fields are horizontal along the horizontal axis. Equations (2.12a, b) to (2.14a, b) also apply to the integrated field parameters such as $C_m{*}L$, $B_m{*}L$, $A_m{*}L$, etc., simply by multiplication of the magnet hard edge length L since the fields are uniform along the length L.

The field components in Cartesian coordinates can be expressed by

$$B_x(r,\theta) = B_r Cos\theta - B_\theta Sin\theta = \sum_{m=1}^{\infty} C_m \left(\frac{r}{R}\right)^{m-1} Sin\left[(m-1)\theta - m\alpha_m\right], \quad (2.15a)$$

$$B_y(r,\theta) = B_r Sin\theta + B_\theta Cos\theta = \sum_{m=1}^{\infty} C_m \left(\frac{r}{R}\right)^{m-1} Cos\left[(m-1)\theta - m\alpha_m\right]. \quad (2.15b)$$

By using the complex variable z = x + iy = r.exp(iθ), Eqs. (2.15a, b) can be combined to obtain a complex field

$$\mathbf{B}(z) = B_y(r,\theta) + iB_x(r,\theta) = \sum_{m=1}^{\infty} C_m \exp(-im\alpha_m)\left(\frac{z}{R}\right)^{m-1} . \tag{2.16}$$

In terms of the normal and skew components defined in Eqs. (2.14a, b), Eq. (2.16) becomes

$$\mathbf{B}(z) = B_y + iB_x = \sum_{m=1}^{\infty} (B_m + iA_m)\left(\frac{z}{R}\right)^{m-1} . \tag{2.17}$$

2.3. Particle Motion

The force on a point charge q in a magnetic field, known as the Lorentz force, is given by

$$\mathbf{F} = q\mathbf{v} \times \mathbf{B} . \tag{2.18}$$

The motion of a charged particle in a magnetic field is then determined by Newton's equation

$$\frac{d\mathbf{P}}{dt} = q\mathbf{v} \times \mathbf{B} , \tag{2.19}$$

where **P** is the mechanical momentum of the particle. The relativistic mechanical momentum **P** is related to the particle velocity **v** by

$$\mathbf{P} = \gamma m\mathbf{v} , \tag{2.20}$$

where γ, also known as the Lorentz factor, is defined as

$$\gamma = \frac{1}{\sqrt{1-\beta^2}} \tag{2.21}$$

and $\beta = v/c$ is the ratio of the particle velocity v to speed of light in vacuum c. The Lorentz force, Eq. (2.18), is determined by the cross product of particle velocity **v** and magnetic flux density **B**. Thus, the magnetic field changes only the direction of particle motion, rather than its energy. The Lorentz factor γ remains a constant. Therefore, the three components in Eq. (2.19) can be written in Cartesian coordinates as

$$\frac{d}{dt}(\gamma m v_x) = \gamma m \ddot{x} = q(\dot{y}B_z - \dot{z}B_y), \tag{2.22a}$$

$$\frac{d}{dt}(\gamma m v_y) = \gamma m \ddot{y} = q(\dot{z}B_x - \dot{x}B_z), \tag{2.22b}$$

$$\frac{d}{dt}(\gamma m v_z) = \gamma m \ddot{z} = q(\dot{x}B_y - \dot{y}B_x). \tag{2.22c}$$

Here the dot on the top of a variable denotes the time derivative of that variable. In a two-dimensional magnetic field as introduced in Section 2.2, the particle dynamics in the transverse planes can be described by the trajectory equations:

$$x'' + \frac{1}{B_\rho} B_y = 0, \tag{2.23a}$$

$$y'' - \frac{1}{B_\rho} B_x = 0. \tag{2.23b}$$

Here the double prime denotes the second derivative with respect to the longitudinal distance z. In deriving Eqs. (2.23a, b) from Eqs. (2.22a, b), we use $z = v_z t$, $v_z \approx v = \text{const}$, and $d^2/dt^2 = v^2(d^2/dz^2)$. B_ρ is the magnetic rigidity of the particle and is defined as the relativistic momentum per unit charge

$$B_\rho = \frac{\gamma m v}{q} \ \text{(T m)}. \tag{2.24a}$$

For proton particles, a useful formula for B_ρ is

$$B_\rho \approx 3.12976(\beta\gamma) \approx 3.33564P\left(\frac{GeV}{c}\right) \text{ (T m)} . \tag{2.24b}$$

At an energy of 1 GeV, one has $\gamma = 2.06578$, $\beta = 0.87502$, $P = 1.69604$ (GeV/c), and $B_\rho = 5.6574$ (T m).

2.4. Dipole Magnets

Dipole magnets have two poles and a uniform field in the gap. They are used to bend charged particle beams in particle accelerators and beam lines. A dipole magnet used for the SNS ring injection chicane is shown in Figure 2.1. This is an H-frame dipole magnet. The two poles are connected by an iron yoke for magnetic flux returns, and they are driven by two current coils with the same polarity. The magnetic flux density B_0 at the pole tip, as well as in the gap between the two poles, is in the y-direction and can be found from the coil current I as

$$B_y = B_0 = \mu_0 \frac{NI}{R} \text{ (T)} , \tag{2.25}$$

where N is the number of turns of the current coil per pole, and $2R$ is the gap distance. Due to the material loss in reality, Eq. (2.25) usually overestimates B_0 by a few percent. The H-frame dipole magnet has a complete symmetry with respect to its three coordinate planes. Thus, ideally, $B_y = B_0 =$ const. is the only component within the gap, and B_x and B_z should vanish. The dipole field in the gap region can also be expressed, in cylindrical coordinates, as

$$B_r(r,\theta) = B_0 Sin(\theta), \tag{2.26a}$$

$$B_\theta(r,\theta) = B_0 Cos(\theta). \tag{2.26b}$$

These equations depict a two-pole field distribution (m = 1, $\alpha_m = 0$) along the circle of radius R. Obviously, this magnetic dipole is a normal dipole and its magnetic fields are vertical along the horizontal axis, as described in the paragraph after Eqs. (2.14a, b).

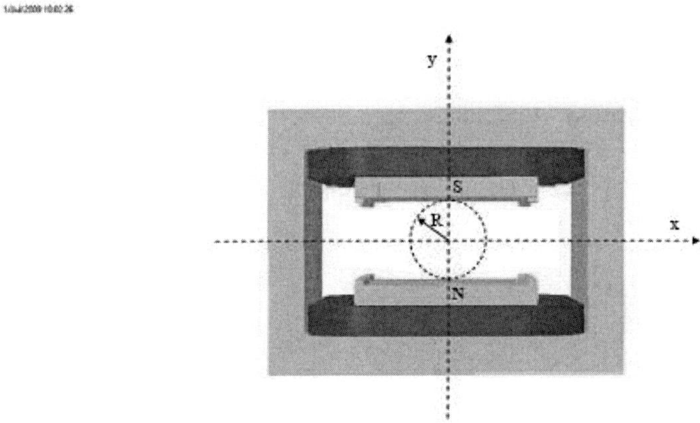

Figure 2.1. A dipole magnet for SNS ring injection.

A charged particle incident into the dipole magnet on the mid-plane, as well as on the other planes parallel to the mid-plane, performs the cyclotron motion in a uniform magnetic field. The particle is thus bent horizontally and stays on the plane. The particle trajectory is curved inside the magnet. The bending angle is determined by

$$\theta = \frac{B_0 L}{B\rho} \text{ (rad)}, \tag{2.27}$$

where L is the curved length inside the magnet. In reality, the dipole field does not end abruptly at magnet ends and B_y is not uniform through the dipole magnet. There is a fringe field which forms a transition from the ideal dipole field B_0 to the field-free region. Thus, the dipole field is a function of s, where s is the distance along the curved particle trajectory. Conventionally, a hard edge approximation can be made by defining an effective dipole length L_{eff} as

$$L_{eff} = \frac{1}{B_0} \int_{-\infty}^{+\infty} B_y(s) ds. \tag{2.28}$$

With the hard edge length L_{eff}, Eq. (2.27) is still valid, and the bending angle of a dipole magnet is determined by its integrated dipole field.

Another widely used dipole magnet configuration is shown in Figure 2.2. This is called a C-shaped dipole magnet. Its open side can be used to insert other components. The C-shaped dipole magnet has no symmetry with respect to the x = 0 plane, and it has a field gradient in the x-direction. Otherwise, their field distributions and particle motions inside the magnet are the same as in the H-framed dipole magnets. There are also other dipole magnets such as septum magnets, which will be described in Chapters 8 and 9.

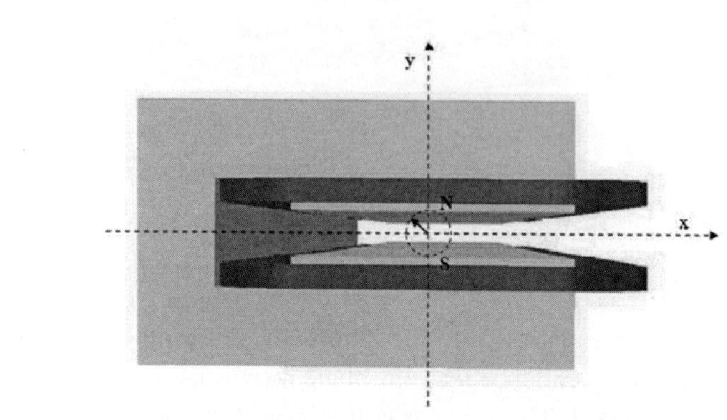

Figure 2.2. A C-shaped dipole magnet for SNS high energy beam transport line.

2.5. Quadrupole Magnets

Quadrupole magnets have four poles and a zero field at their center. They are employed in particle accelerators and beam lines to focus or defocus charged particle beams. The four magnetic poles in a quadrupole magnet are positioned symmetrically around a circle of radius R with the circle center on the axis of the beam line. This is illustrated by an electromagnetic quadrupole for the SNS ring, as shown in Figure 2.3. The four poles have hyperbolic shapes in the transverse cross section. They are driven by four direct current coils with alternative polarities, leading to the poles distributed as N, S, N, S around the circle. This is a normal quadrupole according to its four pole distributions, which result in a vertical field on the x-axis. Magnetic quadrupoles have a linear field in the radial direction. For a two-dimensional model, the gradient g_0 of this linear field is determined by

$$g_0 = 2\mu_0 \frac{NI}{R^2} \text{ (T/m).} \tag{2.29}$$

As in Eq. (2.25), NI is the Ampere-turns of the conducting coil per pole. Equation (2.29) also slightly overestimates g_0 in practice due to the material loss. This equation is often used as a first check of quadrupole simulation or measurement results.

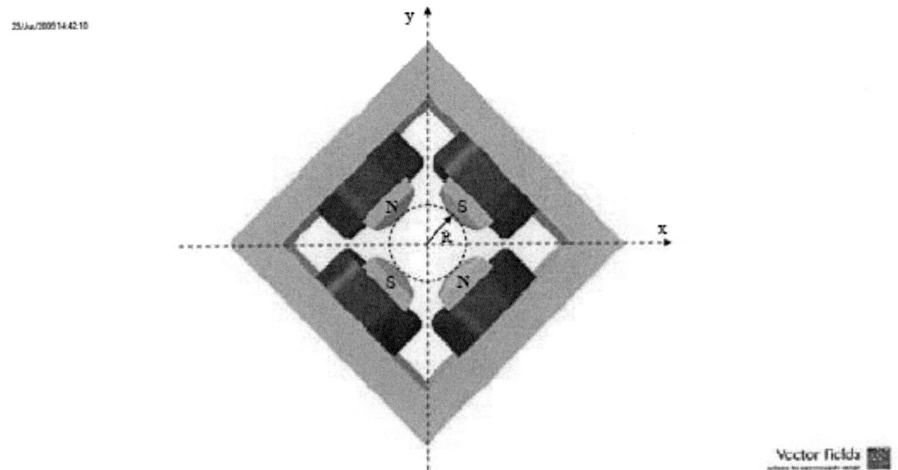

Figure 2.3. An SNS ring quadrupole magnet.

The poles in the quadrupole magnet are distributed around the circle in a way to produce a magnetic quadrupole field described, in cylindrical coordinates, by

$$B_r(r,\theta) = B_0 \frac{r}{R} Sin(2\theta), \tag{2.30a}$$

$$B_\theta(r,\theta) = B_0 \frac{r}{R} Cos(2\theta), \tag{2.30b}$$

or, in Cartesian coordinates, by

$$B_x = B_0 \frac{y}{R}, \tag{2.31a}$$

$$B_y = B_0 \frac{x}{R}. \quad (2.31b)$$

Here B_0 is the magnetic flux density at the pole tip. Along the horizontal axis (y = 0) of this quadrupole, the magnetic fields are vertical (B_y only). It is also clear that the skew term $A_2 = 0$ in Eqs. (2.30a, b), and this is a normal quadrupole with m = 2 and $\alpha_m = 0$.

The linear trajectory equations for a charged particle in the quadrupole magnet can be obtained by plugging the field equations (2.31a, b) into the trajectory equations (2.23a, b) as

$$x'' + kx = 0, \quad (2.32a)$$

$$y'' - ky = 0, \quad (2.32b)$$

where

$$k = \frac{qB_0}{\gamma mvR} = \frac{g_0}{B_\rho} \quad (2.33)$$

is a focusing constant. The solutions of Eq. (2.32a, b) are

$$\begin{pmatrix} x_2 \\ x_2' \end{pmatrix} = \mathbf{M}_x \begin{pmatrix} x_1 \\ x_1' \end{pmatrix} = \begin{bmatrix} Cos\left(\sqrt{k}L\right) & \frac{1}{\sqrt{k}} Sin\left(\sqrt{k}L\right) \\ -\sqrt{k}Sin\left(\sqrt{k}L\right) & Cos\left(\sqrt{k}L\right) \end{bmatrix} \begin{pmatrix} x_1 \\ x_1' \end{pmatrix}, \quad (2.34a)$$

$$\begin{pmatrix} y_2 \\ y_2' \end{pmatrix} = \mathbf{M}_y \begin{pmatrix} y_1 \\ y_1' \end{pmatrix} = \begin{bmatrix} Cosh\left(\sqrt{k}L\right) & \frac{1}{\sqrt{k}} Sinh\left(\sqrt{k}L\right) \\ \sqrt{k}Sinh\left(\sqrt{k}L\right) & Cosh\left(\sqrt{k}L\right) \end{bmatrix} \begin{pmatrix} y_1 \\ y_1' \end{pmatrix}. \quad (2.34b)$$

Here (x_1, x_1') and (x_2, x_2') denote the particle position and slope at the magnet entrance and exit, respectively, and L is the magnet length. $\mathbf{M}_x$ and $\mathbf{M}_y$ are the first-order transfer matrix, and are often called as **R**-matrix in literature. The determinant of the first-order transfer matrix is exactly unity. It is easy to see that the x-motion is focusing and the y-motion is defocusing. The focusing length f_x or f_y of a quadrupole magnet is given by

$$f = -\frac{1}{M_{21}} \ . \tag{2.35}$$

Note that f is positive for focusing, and is negative for defocusing.

In reality, all quadrupole magnets have finite length and the field has fringes. As in the case of dipole magnets, one conventionally defines an effective quadrupole length L_{eff} by

$$L_{eff} = \frac{1}{g_0} \int_{-\infty}^{+\infty} g(z) dz \ , \tag{2.36}$$

where $g(z)$ is the z-dependent gradient, g_0 is the gradient in the central region of the quadrupole magnet.

A skew quadrupole magnet is one that is rotated by 45° from a normal quadrupole, as shown in Figure 2.4. Its field distribution is expressed, in cylindrical coordinates, as

$$B_r(r,\theta) = B_0 \frac{r}{R} Sin\left[2\left(\theta - 45^o\right)\right], \tag{2.37a}$$

$$B_\theta(r,\theta) = B_0 \frac{r}{R} Cos\left[2\left(\theta - 45^o\right)\right], \tag{2.37b}$$

or, in Cartesian coordinates, as

$$B_x = -B_0 \frac{x}{R}, \tag{2.38a}$$

$$B_y = B_0 \frac{y}{R}. \tag{2.38b}$$

In this case, the magnetic fields are horizontal along the horizontal axis. By plugging Eqs. (2.38a, b) into the trajectory equations (2.23a, b), one obtains

$$x'' + ky = 0 \ , \tag{2.39a}$$

$$y'' + kx = 0 \ . \tag{2.39b}$$

In comparison with Eqs. (2.32a, b) for the normal quads, Eq. (2.39a) and Eq. (2.39b) are coupled second-order differential equations. The consequence of the coupling across the x and y planes will be demonstrated in Chapter 9.

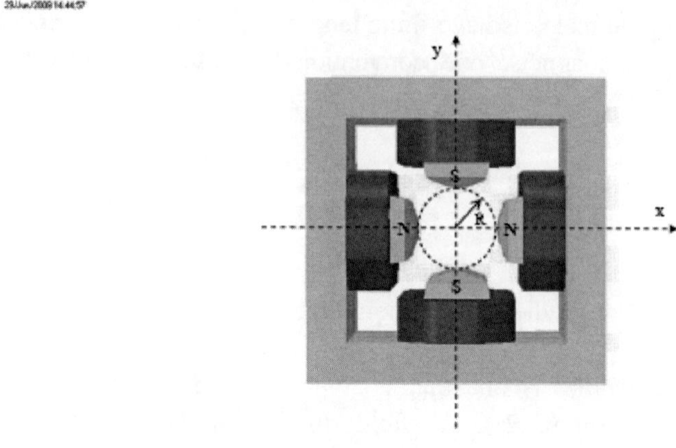

Figure 2.4. A skew quadrupole magnet.

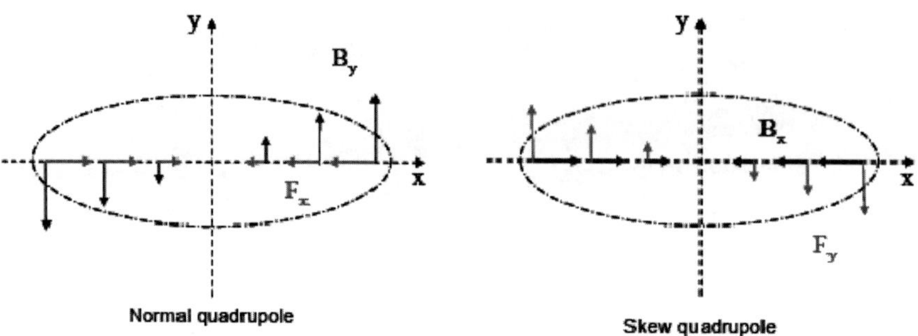

Figure 2.5. Magnetic forces on proton particles along the x-axis from a normal and skew quad.

Figure 2.5 illustrates the difference between a normal and a skew quadrupole in terms of the forces on proton particles along the x-axis, where the particles move in the positive z-direction (come out of paper). The normal quadrupole focuses the particles on the x-axis, while the skew quadrupole gives the particles a twist on the x-axis. On the y-axis, it is easy to see that the normal quadrupole defocuses the particles, while the skew quadrupole twists the particles in the counter-clockwise direction. In fact, the skew quadrupole focuses and defocuses a particle beam on two planes at $+45^{o}$ and -45° with respect to the y = 0 plane. Thus, a skew quadrupole component would distort the beam profile in a lattice consisting of all the normal quadrupoles.

Two quadrupole lenses, arranged as a focusing-defocusing pair, have a net focusing effect in both the x and y directions. This arrangement is the so-called magnetic quadrupole doublet, and is widely used in accelerators and beam lines. If two quadrupole lenses have a focal length f_1 and f_2, respectively, and they are apart by a distance s, the combined quadrupole doublet is equivalent to that of a single lens of focal length F given by

$$\frac{1}{F}=\frac{1}{f_1}+\frac{1}{f_2}-\frac{s}{f_1 f_2}. \qquad (2.40)$$

The particle optics in an SNS quadrupole doublet assembly will be described in details in Chapter 7.

2.6. Magnet Measurements

Electromagnets in particle accelerators are used as beam guiding elements and require very tight tolerance on their magnetic fields along the particle path. Magnetic measurements of accelerator magnets are very important to assure their quality for use in accelerators [5, 7]. In addition, good measurement data of magnets can be used to derive correct particle optics in the devices and to guide machine operations. There are various methods for magnetic field measurements such as the Hall-probe measurements, fluxgate magnetometer, induction coil measurements, etc. Comprehensive review articles on magnet measurement methods can be found in references [8, 9]. For accelerator magnets, the integrated fields and higher-order harmonics are important parameters in applications. These parameters can be obtained at very high accuracy by measurements of magnets with induction coils, or so called search coils [10, 11]. There are basically two kinds of search coil configurations: the tangential coil and radial coil. The SNS magnets were measured at ORNL by various radial coils.

A radial search coil consists of a conducting winding on a plane through the magnet axis. The two sides of the winding are in parallel with the magnet axis and are located at radial positions r_1 and r_2. The coil is long enough to cover all the fringe field regions. It rotates about the magnet axis. The magnetic flux through the coil is an integration of the azimuthal field component B_θ over the winding area at a given rotation angle. By using Eq. (2.12b) for B_θ, the magnetic flux through the coil can be obtained as

$$\Phi(\theta) = N\int_{-\infty}^{+\infty} dz \int_{r_1}^{r_2} dr B_\theta(r,\theta) = \sum_{m=1}^{\infty} (C_m * L) Cos[m(\theta - \alpha_m)] \frac{1}{mR^{m-1}} N(r_2^m - r_1^m), \tag{2.41}$$

where N is the number of turns of the coil winding, and (C_m*L) represents the integrated field amplitude over the entire length of the coil. When the coil rotates, the magnetic flux $\Phi(\theta)$ changes with the angle θ, or time t. Thus, a voltage signal is produced at the output of the coil according to

$$V(t) = -\frac{d\Phi(t)}{dt}. \tag{2.42}$$

This signal is fed into an electronic integrator, which recovers the magnetic flux $\Phi(\theta)$. By performing Fourier analysis of $\Phi(\theta)$

$$\mathcal{F}[\Phi(\theta)] = \sum_{k=1}^{\infty} X_k \angle \theta_k, \tag{2.43}$$

we find the integrated field amplitude and phase angle for each harmonic number $m = k$ as

$$C_m * L = X_k (mR^{m-1}) / N(r_2^m - r_1^m), \tag{2.44a}$$

$$\alpha_m = \theta_k / m. \tag{2.44b}$$

The integrated normal and skew terms can be found by Eqs. (2.14a, b). For magnetic dipoles, it is interesting to know the integrated bending field ($B*L = C_1*L$) and the integrated harmonics (C_m*L, $m>1$). It is a common practice to express the normalized coefficient NC_m of the integrated harmonic content as

$$NC_m = \frac{(C_m * L)}{(C_1 * L)} \bullet 10^4 = \frac{(C_m * L)}{(B * L)} \bullet 10^4 \text{ (units)}. \tag{2.45}$$

There is a similar formula for magnetic quadrupoles, where the denominator is replaced by C_2*L.

For accelerator magnets, the main integrated field such as C_1*L in a dipole magnet and C_2*L in a quadrupole magnet can be accurately measured by a single, radial coil. However, this is not the case for the higher harmonics since they are usually very small at a level of 10^{-3} to 10^{-4} of the main fields. It is difficult to accurately pick up such small signals from the much stronger main field signal. The remedy for the problem is to use the so-called Halbach coil [12], which consists of the main winding, as described above, and a secondary winding or bucking winding. Halbach proposed and showed that if a secondary winding is located at radial positions r_3 and r_4 on the same plane as for the primary winding to satisfy the following conditions

$$r_2 - r_1 = a(r_4 - r_3), \tag{2.46a}$$

$$r_2 + r_1 = (r_4 + r_3), \tag{2.46b}$$

where a is the ratio of the numbers of turns between the two windings, the magnetic flux from the two windings would contain the same amount of m = 1 and m = 2 terms. Thus, it is possible to connect the two windings in such a way that the net output contains no main field signal (m = 1 and m = 2). This is so called bucking, which can greatly improve the measurement accuracy of the higher harmonic components.

Figure 2.6 shows a block diagram of the SNS magnet measurement setup. All the search coils are the Halbach type with a = 1. For the primary winding, r_2 is usually located at about 80% of the magnet aperture, while r_1 is about half the value of r_2 but on the opposite side of the magnet axis (r_2 and r_1 have opposite signs). The bucking winding dimensions are determined by Eqs. (2.46a, b). The electronic equipment includes a digital integrator, a digital volt meter, a current monitor, and power supplies. A LabVIEW program [13] is used to control the measurement and to process the data automatically. Many details of measurements, such as the accuracy and digitizing errors, can be found in [14].

The method of magnet measurements with the radial coils can also be applied in simulations. In magnet simulation models, we create a rectangular patch, which is similar to the primary winding with N=1 in a Halbach coil. The bucking winding is not needed since there are no noise signals as in measurements. The magnetic flux through the patch can be computed in the same way as shown by Eq. (2.41), and subsequent analyses remain the same. If the patch dimension is the same as in the primary coil, the patch rotation method in simulations can be directly compared with experiments. Figure 2.7 shows the patch rotation method applied in an SNS quadrupole [15]. The rectangular patch is set at different angular positions for magnetic flux calculations. This method will be frequently used in subsequent chapters of this book.

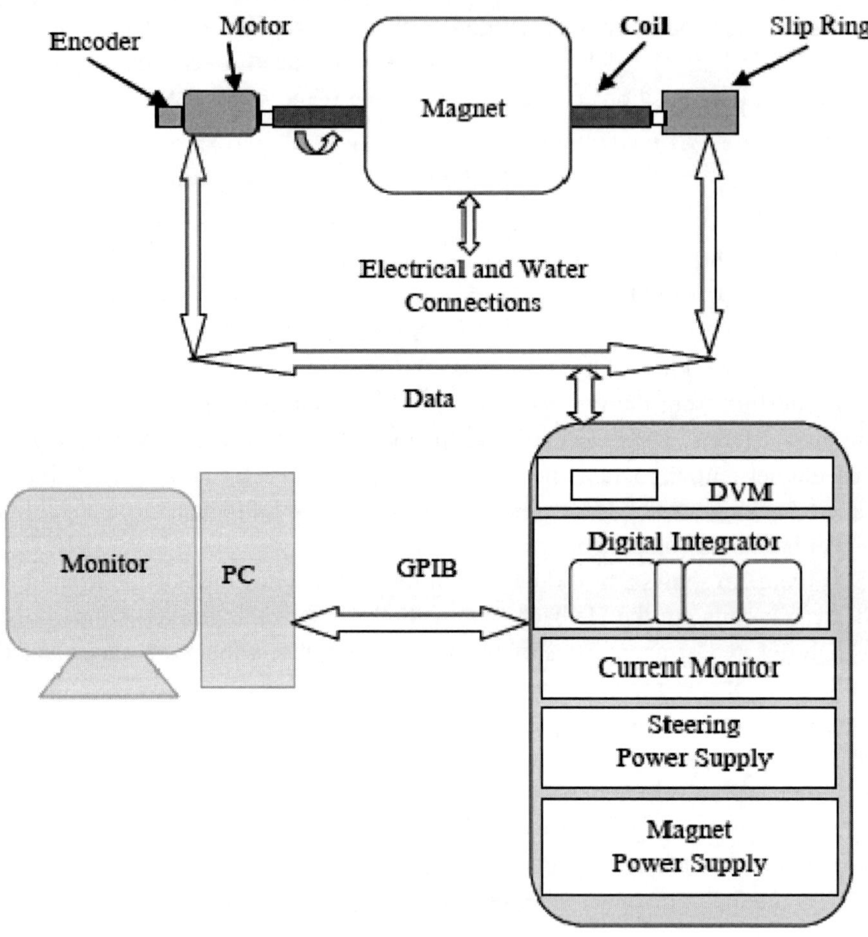

Figure 2.6. A block diagram of the SNS magnet measurement setup.

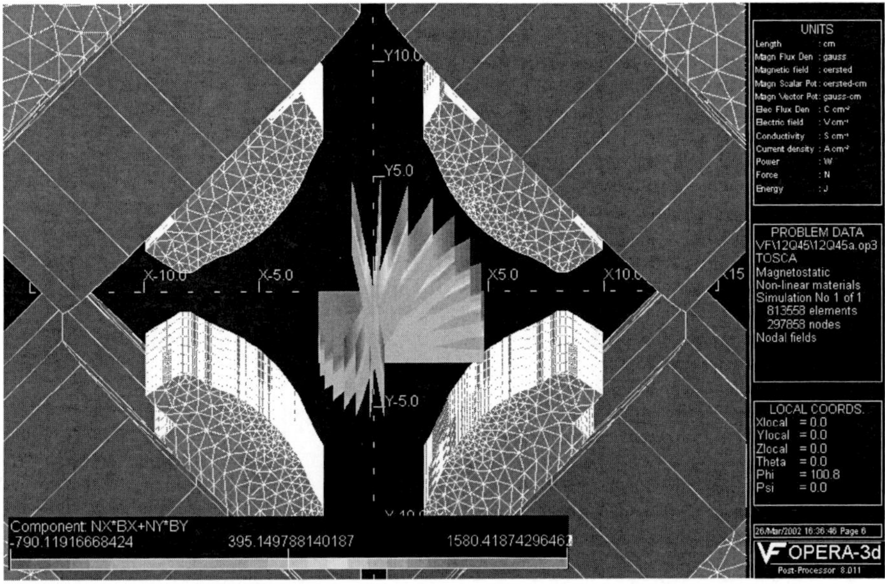

Figure 2.7. A rectangular patch rotates in an SNS quadrupole in simulation.

REFERENCES

[1] D. K. Cheng, *Field and Wave Electromagnetics,* Addison-Wesley Publishing Company, New York, 2nd Ed., 1992.

[2] R. K. Wangsness, *Electromagnetic Fields*, John Wiley and Sons, New York, 1979.

[3] M. Reiser, *Theory and Design of Charged Particle Beams*, John Wiley and Sons, Inc., New York, 1994.

[4] J. D. Lawson, *The Physics of Charged-Particle Beams*, Clarendon Press, Oxford, 2nd Ed., 1988.

[5] J. T. Tanabe, *Iron Dominated Electromagnets: Design, Fabrication, Assembly and measurements*, World Scientific, Singapore, May 2005; also see SLAC-R-754, June 2005.

[6] A. K. Jain, "Basic theory of magnets," in Proceedings of CERN Accelerator School on Measurement and Alignment of Accelerator and Detector Magnets, Anacapri, Italy, April 11–17, 1997, ed. S. Turner (CERN, Geneva, 1998), p.1.

[7] E. J. N. Wilson, "The motivation for magnet measurements," ibid, p. 27.

[8] K. N. Henrichsen, "Overview of magnet measurement methods," ibid, p. 127.

[9] C. Germain, "Bibliographical review of the methods of measuring magnetic fields," *Nucl. Instrum. Methods*, 21 17 (1963).

[10] W. G. Davies, "The theory of the measurement of magnetic multipole fields with rotating coil magnetometers," *Nucl. Instrum. Methods*, A311 399 (1992).

[11] A. K. Jain, "Harmonic coils," in Proceedings of CERN Accelerator School on Measurement and Alignment of Accelerator and Detector Magnets, Anacapri, Italy, April 11–17, 1997, ed. S. Turner (CERN, Geneva, 1998).

[12] K. Halbach, "The Hilac quadrupole measurement equipment," Engineering Note, LBL, March 3, 1972.

[13] LabVIEW is a graphical programming language developed by National Instruments Corp, Austin, TX.

[14] J. G. Wang and D. Barlow, "Elimination of digitizing errors in a rotating coil mapper," in Proceedings of the Eighth European Particle Accelerator Conference, Paris, France, June 3–7, 2002, eds. T. Garvey et al. (EPS-IGA and CERN, 2002), p. 2385.

[15] J. G. Wang, "Acceptance tests of the SNS transfer line quadrupoles (12Q45)," SNS-NOTE-MAG-112, August 22, 2003.

Chapter 3

MAGNET MODELING

3.1. OVERVIEW OF SIMULATION CODES

The studies of the effect of magnetic fringe fields and interference in accelerator magnets rely on realistic, accurate information on magnetic field distributions in these devices. The only effective technique nowadays to accomplish this task is computer modeling. Magnetic measurements in accelerator labs based on various search coils usually provide two-dimensional field parameters. Although three-dimensional field point mappers do exist, their applications are time-consuming and their data are not in the format required by further particle optics analysis. Nevertheless, measurements are always crucial in verifying simulation accuracy.

There are many computer codes for accelerator magnet design and analysis. A compendium of such codes back to 1990s can be found in reference [1]. For the theories behind these codes, readers may confer early papers such as those contained in references [2, 3]. A more recent review article on computational electromagnetics can be found in [4], where a comprehensive list of references on the subject is provided. Over time, some of the codes in [1] have become obsolete, while others become widely distributed, and new codes are developed.

In recent years, the accelerator magnet community conducted three special workshops on magnet simulations for particle accelerators [5–7]. The workshops were intended to bring magnet physicists and engineers together with simulation code developers to exchange ideas and to discuss commonly interested topics in this special field. The workshops were attended by professionals from national laboratories, private industries, and other academic and research institutions all over the world. The participants used this platform to discuss the applications of magnet design codes such as ANSYS/Maxwell [8], COMSOL [9], MagNet [10],

MAFIA [11], MERMAID [12, 13], OPERA/TOSCA [14, 15], POISSON [16–18], RADIA [19], ROXIE [20], and etc. The workshops covered the modeling of various accelerator magnets, including iron-dominated electromagnets, superconducting magnets, permanent magnets, insertion devices such as wigglers and undulators, and detector magnets. Other interesting and useful topics were exampled by the construction of accurate particle optics from magnetic field data or particle trajectory data in magnet modeling. The workshop discussions greatly benefited the study of magnetic fringe fields and interference in high-intensity accelerators.

Electromagnetism is governed by Maxwell's equations. For magnetostatic field computation, Maxwell's equations are reduced to Poisson's equation (Eq. (2.5)) or Laplace's equation (Eq. (2.9)) with the introduction of either a vector or a scalar magnetic potential. All these equations are Partial Differential Equations (PDEs). With the exception of a few simplest cases where the analytical solutions are available, most accelerator magnet problems involve complicated geometries, which require numerical solutions of the PDEs with specified boundary conditions. Thus, any simulation code for magnetic design and analysis should accomplish four tasks: to build a geometric model for a real magnet; to discretize the subject domain with the so-called meshes; to solve the governing PDEs with a mathematic method; and to post-process the results in order to obtain magnetic field distributions, other derived quantities, particle trajectories, and etc.

Accelerator magnets usually consist of magnetic materials (soft irons, permanent magnets, etc) in various geometric shapes, current-carrying conductors (room temperature or superconducting), and surrounding air environment. Modern accelerator magnet design codes provide a suite of facilities for building simulation models. This includes the primitives of geometric definitions, Boolean operations, and the ability to import and export data from most popular CAD systems. Once a simulation model is built with the user defined parameters such as cell properties, mesh sizes, boundary conditions, and etc., the mesh generation is automatically accomplished in design codes. There are various mesh geometries in different codes, depending on the mathematical method a code uses to solve the PDEs.

The classical solution method for numerically solving PDE's is the Finite Difference Method (FDM) [21], in which the differential operators are approximated by difference operators based on a grid built up from coordinate planes. The method faces difficulties in anisotropic materials, on irregular material boundaries, and for generating symmetric matrices. Thus, the Finite Difference Method is not popular nowadays in solving magnetic field problems. The Finite Element Method (FEM) [22, 23] is an advanced numerical technique for finding approximate solutions of PDEs. It is most commonly used in many accelerator magnet design codes such as ANSYS/Maxwell, COMSOL, MagNet, POISSON, and TOSCA. In the Finite Element Method, the subject domain (iron, conductor, air) is first divided into elementary volumes, i. e., the “finite elements”. The variational integral is then approximated numerically on each of these elements in terms of the node values of the potential. The contributions to the total integral are summed over all elements, and the magnetic field is derived from the potential by numerical differentiation. The main advantages of the FEM are the ease of modeling complicated boundaries and the extendibility to higher order approximations. In addition to the differential methods, integral methods are also used in accelerator magnet design codes, such as the finite volume integral method in RADIA and the boundary integral method in MAFIA. The integral methods [24, 25] avoid some difficulties of defining boundary conditions in the subject domain with unlimited regions. Only the iron and the conductors are specified and there is no mesh in the surrounding air. These methods are more advantageous for the problems with simple geometry and linear material properties.

Post-processors are an essential constituent of all simulation codes. The solutions from the solvers have to be interpreted and displayed for the physical parameters of designed magnets. A suite of facilities is usually provided to enhance magnetic field calculations at points, along lines, on two-dimensional patches, and for other derived quantities and particle trajectories in simulated fields. Many codes provide state-of-the-art Graphical User Interface (GUI) to accomplish these tasks.

Although many good accelerator design codes exist, OPERA/TOSCA is probably the most widely used for accelerator magnet design in national accelerator labs and other institutions. Because the author is more familiar with this code, we thus describe it in more details below. This should not be taken as an endorsement of a commercial product. In this discussion, we limit ourselves to magnetostatic cases.

3.2. OPERA/TOSCA

OPERA stands for **OP**erating environment for **E**lectromagnetic **R**esearch and **A**nalysis. It consists of the pre and post processing system for electromagnetic analysis programs such as TOSCA for nonlinear magnetostatic or electrostatic field and current flow problems, ELEKTRA for time varying eddy current and transient problems, SCALA for three-dimensional space charge problems,

SOPRANO and CONCERTO for high frequency and microwave design problems. The codes are developed by Vector Fields Software of Cobham Technical Services in England.

OPERA offers both two-dimensional and three-dimensional packages. OPERA-2D is a suite of programs for two-dimensional electromagnetic field analysis. It solves models with axial symmetry or XY- symmetry at high accuracy with least simulation complexity. Although it is also widely employed in the initial design and optimization of real three-dimensional magnets, it is usually not adequate to solve problems involving magnetic fringe fields and interference, which require true three-dimensional treatments.

OPERA-3D employs two modules, the MODELLER and the PRE PROCESSOR, to generate geometric models and data files for the electromagnetic analysis programs such as TOSCA. Both modules are interactive programs that allow the user to create and edit three dimensional finite element mesh, define material characteristics including nonlinear and anisotropic description, assign boundary conditions, specify complicated conductor geometries with prescribed excitation, and graphically display and examine the data. In the OPERA-3D PRE PROCESSOR, the model is built by using a baseplane and extrusion method. A suitable two-dimensional cross-section of a magnet is drawn on a baseplane by coordinate points and lines to form facets, which are further divided to a two-dimensional finite element mesh. The two-dimensional cross section is then extruded in the third dimension to build a three-dimensional structure. This method faces difficulties for complicated geometries. The MODELLER uses geometric primitive volumes and Boolean operations to construct simulation models. It manipulates any defined objects through operations such as transformation and combination. Basic objects (blocks, cylinders, spheres, cones, pyramids and toroids) can be created at any position in space. Once created, they can be manipulated to reposition themselves. They can also be merged, intersected or subtracted from other objects in space to create more complex geometries. Other more advanced techniques allow the geometry to be enhanced by techniques such as sweeping an existing face. The MODELLER can also import and export data files from most popular CAD systems. The MODELLER is very powerful in building complicated geometries and is especially suited for modeling many magnets together.

TOSCA is one of the three-dimensional electromagnetic analysis programs offered in OPERA. The package solves magnetostatic or electrostatic field and current flow problems by state-of-the-art finite-element techniques. A typical magnet model can be divided into three different regions, as shown in Figure 3.1. The region having non-zero conductivity and permeability is denoted by Ω_C. This region would include ferromagnetic materials such as iron ("soft" materials) and permanent magnets ("hard" materials). The region called Ω_S contains field sources (i.e. current carrying coils). These two regions are usually surrounded by the region Ω_0, which is the air (i.e. free space). For numerical modeling by the Finite Element Method, the problem must be bounded. A suitable distant boundary is denoted by Γ_0.

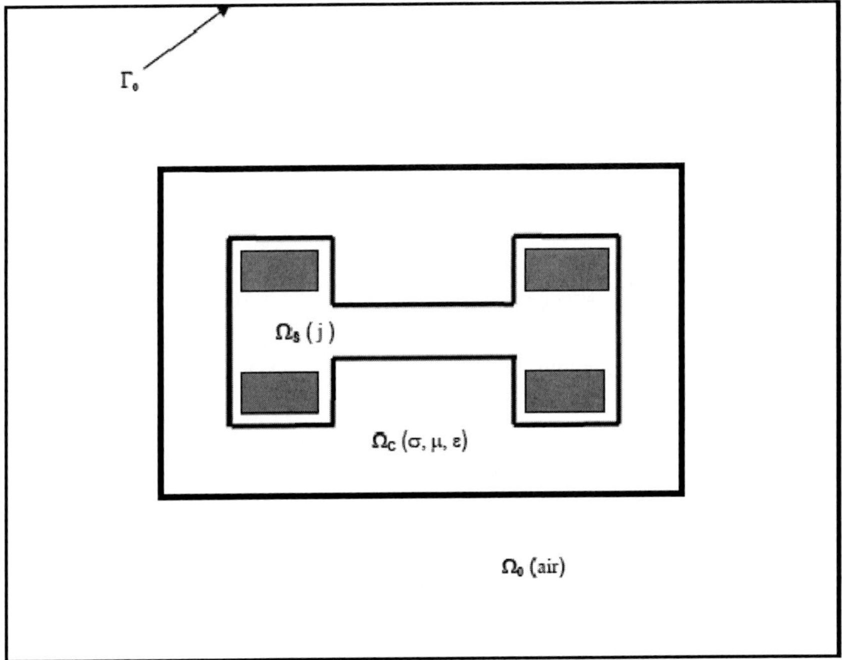

Figure 3.1. Different regions in the model space.

In a current-free region such as Ω_c or Ω_0, the magnetic flux intensity **B** is curl-free (Eq. (2.1b)) and can be expressed as the gradient of a magnetic scalar potential Φ_m such that

$$\mathbf{B} = -\mu \nabla \Phi_m , \tag{3.1}$$

where the magnetic scalar potential Φ_m is called the total scalar potential in TOSCA. In a three-dimensional region such as Ω_S where predefined source currents exist, a scalar potential can still be defined. This is done by partitioning the magnetic field intensity **H** into two parts:

$$\mathbf{H} = \mathbf{H}_s + \mathbf{H}_m, \tag{3.2}$$

where $\mathbf{H_s}$ is the magnetic field intensity due to source current density **J**, and $\mathbf{H_m}$ is the magnetic field intensity due to material magnetization. The former can be obtained from the Biot-Savart law over a volume Ω_s containing currents. The later, which is now curl-free, can be related to a magnetic scalar potential ϕ as

$$\mathbf{H}_m = -\nabla\phi. \tag{3.3}$$

This yields an expression for the total magnetic field intensity of

$$\mathbf{H} = \mathbf{H}_s - \nabla\phi. \tag{3.4}$$

The scalar potential ϕ is called the reduced scalar potential in TOSCA. Equation (3.4) shows that finding ϕ and knowing $\mathbf{H}_s$ allow the determination of **H** and **B**.

The TOSCA method computes the total scalar potential in the magnetic material and the reduced scalar potential in the region where source currents have been specified. The reduced scalar potential represents only that portion of the field produced by magnetization, and the remainder of the field is computed directly by source currents. By using this method, TOSCA avoids the drawbacks of other methods that often produce cancellation errors. The code is well known for its high accuracy as proven by many years of comparison with measured results. The magnetic material properties for TOSCA may be specified as nonlinear and anisotropic. Permanent magnets can be represented with a specified magnetization direction or the magnetization distribution may be predicted by simulation. The program uses an iterative solution technique for the matrix of linear simultaneous equations obtained for the potential at each node of the mesh, thereby reducing considerably the memory requirements needed for a direct matrix solution algorithm. The TOSCA program employs a modified Newton-Raphson technique to successfully update the element permeability in order to obtain the fields with nonlinear materials present.

The OPERA-3D POST PROCESSOR displays and performs further calculations on results from electromagnetic field analysis programs including TOSCA and many others. The POST PROCESSOR provides facilities to view the finite element data, to process and display the electromagnetic field at points, along lines, and on two dimensional areas. It can also calculate many derived quantities such as field integrals, body forces, etc., compute and plot particle trajectories through calculated fields.

3.3. A Modeling Example

As an example of OPERA-3D/TOSCA application, we present the modeling of two chicane dipoles in the SNS ring injection region [26]. The result will be used for three-dimensional multipole expansion in the next chapter. Figure 3.2 shows a model of chicane dipoles D2 and D3. They were designed and developed at Brookhaven National Lab (BNL) [27]. Both dipoles are the C-shaped dipoles with specially designed complimentary pole tips. The two dipoles have a nominal gap height of 24.7904 cm, pole width of 50.038 cm, and 14 turns per coil. The nominal pole length is 70.104 cm for D2 and 47.9552 cm for D3. The distance between the D2 gap center and the D3 gap center is 181.4068 cm. These two large aperture dipoles are so close that their fringe fields overlap. The magnetic fringe fields and interference between the two dipoles can be taken care of automatically in a single simulation model containing the two magnets together. The origin of the coordinates in the simulation model is located at the D2 gap center. In the model, dipoles D2 and D3 are energized at 2140 A and 1690 A, respectively. These currents should be close to the operation values for a 1 GeV proton beam. According to Eq. (2.25), the magnetic flux density B_0, which is in the y-direction in the central D2 and D3 regions, should be 3037.4 and 2398.7 G, respectively. These numbers can be used as a first check of the simulation results.

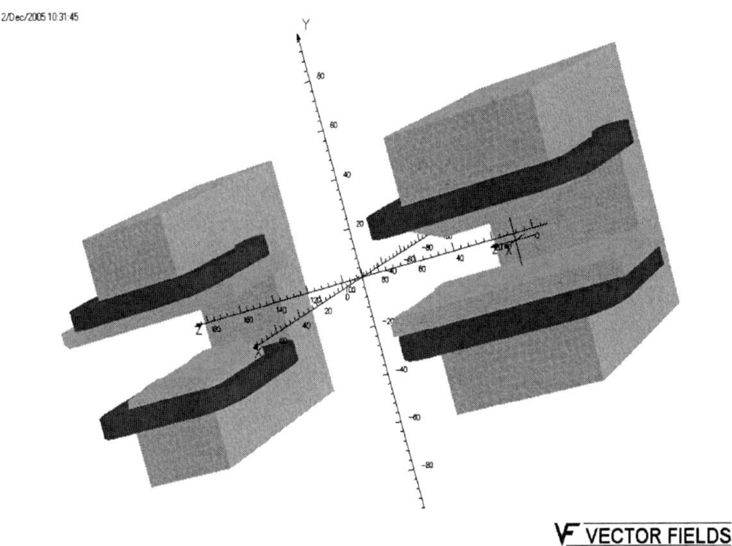

Figure 3.2. Three-dimensional simulation model of SNS injection chicane dipoles D2 (right) and D3.

The simulation model is built with OPERA-3D MODELLER. We first simulate the two dipoles separately, and then combine them together. In the geometrical model we create an air cell surrounding each conducting coil. This cell is defined as a reduced scalar potential region. The other regions are specified as total scalar potential regions. Since there are no symmetry planes for the two magnets together, we have to mesh the whole space. A very large background cell is drawn to enclose the two dipoles. This results in difficulties in volume mesh and requires large computer memory. An artificial cut of a plane in the middle of the two dipoles alleviates the problem. After all mesh sizes, cell properties, boundary conditions, etc. are defined, the program completes surface mesh and volume mesh automatically and creates a database file for solution. The magnetostatic solver TOSCA is then launched to compute the magnetic fields and to store data in a post-processor file, from which we obtain all the desired field quantities in OPERA-3D POST PROCESSOR.

The magnetic field on the z-axis from the simulation is plotted in Figure 3.3. The dominant component is the dipole field B_y, which has a minimum at $z = 93.87$ cm between D2 and D3. The peak value of B_y is 3009.36 G at $z = 0$ for D2 and 2353.59 G at $z = 181.4$ cm for D3. These values are 0.93% and 1.9% smaller than those predicted by Eq. (2.25). The axial field B_z is also rather large, as shown by the dashed red curve. This is an indication of significant fringe fields produced by the dipoles. The integrated field and the integrated harmonic contents are analyzed by a rotating Cartesian patch, similar to an un-bucked winding in a Halbach search coil in accelerator magnet measurements as explained in Section 2.6. Figure 3.4 shows the integrated harmonic distribution evaluated within a reference radius of 8 cm. The integrated harmonic amplitudes are expressed in units with one unit as 1E-4 of the integrated dipole field. Note that the vertical axis is in logarithmic scale. The blue and green bars are calculated for the D2 region and D3 region respectively with a separation point at $z = 93.87$ cm. The red bars are computed for the entire length of both D2 and D3. It is clear that the overall integrated strength of the quadrupole, sextupole, and octupole terms through the two dipoles is much smaller than that in each individual dipole. This is due to the opposite phases of these harmonics in the two dipoles as produced by complementary pole tips.

For beam injection into the SNS ring, there is a stripping foil, which was initially located at $x = 0$, $y = 2$, and $z = 30.7$ cm in the original design (note that the origin of the coordinates is at the center of the D2 gap). This location is around the left edge of D2 in Figure 3.2. It is important to know the field in the stripping foil area since the motion of stripped electrons is determined by the magnetic field lines in that region. In Table 3.1 we list these parameters and compare the simulation results between ORNL and BNL [28]. The agreement looks good.

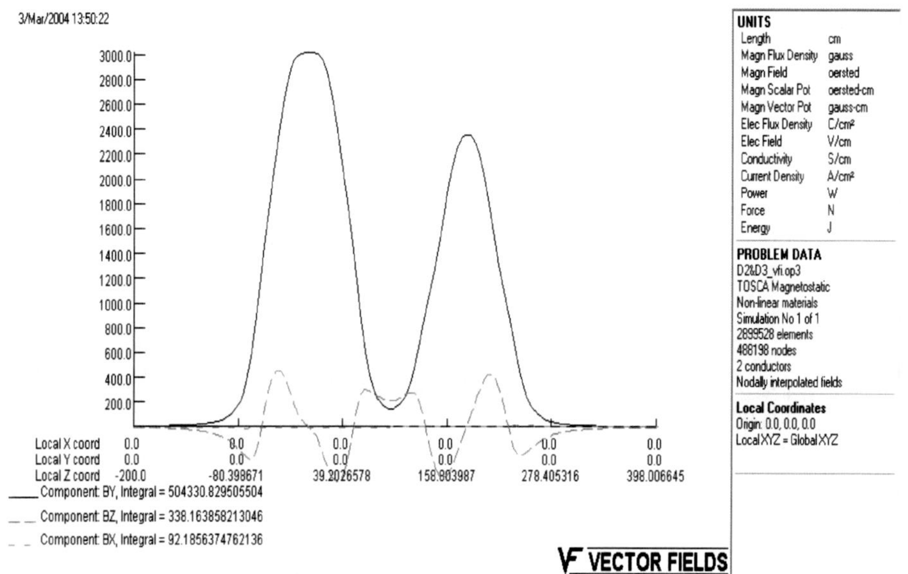

Figure 3.3. Magnetic field distribution along the z-axis.

Table 3.1. Field distribution around stripping foil from simulations

	ORNL simulation	BNL simulation
D2 Current (A)	2140	2168
D3 Current (A)	1690	1716
B_y (kG)	2.5244	2.50
B_z (kG)	-0.5256	-0.532
B_{total} (kG)	2.5785	2.556
$Tan^{-1}(B_z/B_y)$ (rad.)	-0.2053	-0.2
B_y integral (G-cm) (from minus infinity to foil)	241464	237997
B_y integral (G-cm) (from foil to plus infinity)	263164	261751

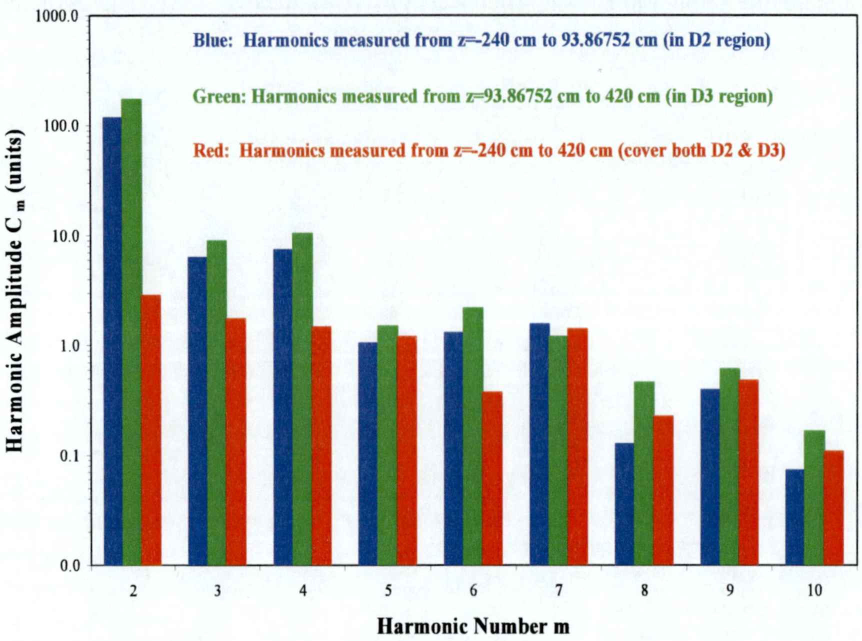

Figure 3.4. Integrated harmonic amplitude of chicane dipoles D2 and D3.

From the D2 and D3 simulation we have generated the magnetic field distribution in three-dimensional grids for further beam tracking studies. This is similar to data generation by a three-dimensional point mapper in experiments. The data file stores the magnetic fields within a rectangular bar, which has a cross section of 20 x 20 cm^2 and a length from z = -200 cm to z = 400 cm. The step size is 0.5 cm in all dimensions. Thus, the file contains more than two million space points, each of which is specified by three Cartesian coordinates (x, y, z) and three field components (B_x, B_y, B_z). The file size is about 250 MB. If a step size of 0.25 cm is desired, it will produce a data file of more than 2 GB. This will be very difficult to handle in subsequent applications. More importantly, the existing frame work of beam optics usually requires the expansion of all the physical parameters including the magnetic field into various orders. The discrete three-dimensional magnetic data on grids do not satisfy this requirement. These issues have motivated us to develop the three-dimensional multipole expansion of magnetic fields from simulation data, as described in the next chapter.

REFERENCES

[1] Los Alamos Accelerator Code Group, *Computer Codes for Particle Accelerator Design and Analysis: A Compendium*, LA-UR-90-1766, Second Edition, May 1990.

[2] C. Iselin, "Review of recent development in magnet computations," *IEEE Trans. on Magnetics*, 17 (5), 2168 (1981).

[3] T. Tortschanoff, "Survey of numerical methods in field calculations," *IEEE Trans. on Magnetics*, 20 (5), 1912 (1984).

[4] C. W. Trowbridge and J. K. Sykulski, "Some key developments in computational electromagnetics and their attribution," *IEEE Trans. on Magnetics*, 42 (4), 503 (2006).

[5] First special workshop on magnet simulations for particle accelerators, organized by J. G. Wang of SNS/ORNL and J. Simkin of Vector Fields, England, in PAC05, Knoxville, TN, May 16–20, 2005. (http://www.sns.gov/pac05/magnet_workshop.shtml)

[6] Second special workshop on magnet simulations for particle accelerators, organized by J. G. Wang of SNS/ORNL and J. Simkin of Vector Fields, England, in PAC07, Albuquerque, NM, June 25–29, 2007. (http://indico.fnal.gov/conferenceDisplay.py?confId=1053)

[7] Third special workshop on magnet simulations for particle accelerators, organized by J. G. Wang of SNS/ORNL and J. Simkin of Vector Fields Software, England, in PAC09, Vancouver, B. C., Canada, May 3–9, 2009. (http://indico.fnal.gov/conferenceDisplay.py?confId=2612)

[8] ANSYS/Maxwell, A low frequency electromagnetic field simulation code by Ansys Inc., see http://www.ansys.com/products/electromagnetics /default.asp

[9] COMSOL, The unifying multiphysics simulation environment by COMSOL Inc., see http://www.comsol.com/

[10] MagNet: 2D/3D electromagnetic field simulation software, Infolytica Corp., Montreal, Canada, see http://www.infolytica.com/

[11] F. Ebeling, R. Klatt, F. Krawzcyk, E. Lawinsky, T. Weiland, S. G. Wipf, B. Steffen, T. Barts, J. Browman, R. K. Cooper, and G. Rodenz, "Status and future of the 3D MAFIA group of codes," in AIP conference proceedings 177: Linear Accelerator and Beam optics codes, La Jolla Institute 1988, ed. C. R. Eminhizer (AIP, N.Y., 1988), p. 117.

[12] A.N. Dubrovin, *Mermaid User's Guide,* SIM Limited, Novosibirsk, Russia.

[13] S. Mikhailov, presentations in the first, second, and third special workshop on magnet simulations for particle accelerators, see references [5–7].

[14] J. Simkin and C. W. Trowbridge, "Three-dimensional nonlinear electromagnetic field computations, using scalar potentials," in IEE proceedings 127 (6), Part B, 368 (1980).

[15] OPERA-3D User Guide and Reference Manual, Vector Fields Software of Cobham Technical Services, Oxford, England;
also see http: //www.vectorfields.com/

[16] K. Halbach, "A program for inversion of system analysis and its application to the design of magnets," Lawrence Livermore National Laboratory report UCRL-17436 (1967); CONF-670705-14.

[17] C. Iselin, POISCR, CERN program library, 1984.

[18] R. C. Gupta, "Improved mesh generator for the POISSON group codes," in Proceedings of the 1987 Particle Accelerator Conference: Accelerator Engineering and Technology, Washington, D. C., March 16–19, 1987, eds. E. R. Lindstrom and L. S. Taylor (IEEE Piscataway, NJ, 1987), p. 1449.

[19] P. Elleaume, O. Chubar, and J. Chavanne, "Computing 3D magnetic fields from insertion devices," in Proceedings of the 1997 Particle Accelerator Conference, Vancouver, B. C., Canada, May 12–16, 1997 (IEEE Piscataway, NJ, 1998), p. 3509. RADIA is free for download from the website: http://www.esrf.eu/Accelerators/Groups/InsertionDevices /Software/Radia

[20] S. Russenschuck, "ROXIE – A computer code for the integrated design of accelerator magnets," in Proceedings of the Sixth European Particle Accelerator Conference, Stockholm, Sweden, June 22–26, 1998, eds. S. Myers et al. (IPP, Bristol, UK, 1998), p. 2017.

[21] For example, K.W. Morton and D.F. Mayers, *Numerical Solution of Partial Differential Equations, An Introduction*. Cambridge University Press, 2005.

[22] G. Strang and G. Fix, *An Analysis of the Finite Element Method*, Englewood Cliffs: Prentice-Hall, 1973.

[23] M. V. K. Chari and P. P. Silvester, Eds., *Finite Elements in Electrical and Magnetic Field Problems*, John Wiley and Sons, New York, 1980.

[24] C. W. Trowbridge, "Applications of integral equation methods for the numerical solution of magnetostatic and eddy current problems," in Proceedings of the International Conference on numerical methods in electrical and magnetic field problems, Santa Margherita, Italy, 1976.

[25] C. W. Trowbridge, "Three-dimensional field computation," *IEEE Trans. on Magnetics*, 18 (1), 293 (1982).

[26] J. G. Wang, "Magnetic field distribution of injection chicane dipoles in Spallation Neutron Source accumulator ring," *Phys. Rev. ST Accel. Beams* 9, 012401 (2006). (http://prst-ab.aps.org/abstract/PRSTAB/v9/i1/e012401); also see J. G. Wang, SNS-NOTE-MAG-130, 2004.

[27] D.T. Abell, Y.Y. Lee, and W. Meng, "Injection into the SNS accumulator ring: minimizing uncontrolled losses and dumping stripped electrons," in Proceedings of the Seventh European Particle Accelerator Conference, Vienna, Austria, June 26–30, 2000, p. 2107.

[28] D. Raparia, "Magnet analysis and engineering support," in presentation to ASAC (Accelerator System Advisory Committee) review of the SNS project, March, 2004.

Chapter 4

THREE-DIMENSIONAL FIELD MULTIPOLE EXPANSION

Magnetic fringe fields and their interference in accelerator magnets can be accurately modeled nowadays via computer simulation. However, the resulting data from computer codes are usually the fields at points, along lines and circles, on two-dimensional patches, and for other derived quantities such as field integrals, etc. None of these field data is in the form required by particle optics analysis. An expansion of the magnetic fields about a reference trajectory to three-dimensional multipoles is a necessary step to further study the topic.

The fundamental theory of the three-dimensional multipole expansion of magnetic fields in cylindrical coordinates has long been established [1]. Some early work on this topic for accelerator magnet analyses and measurements can be found in references [2–7]. More recent and comprehensive treatments of the subject were systematically developed in references [8, 9]. The techniques are based on magnetic field data obtained through measurement or computation. The expansion is not only for the magnetic scalar potential, but also for the magnetic vector potential. Realistic field data on surfaces of various geometries are studied in reference [10].

Following the material presented in [8, 9, 11, 12], we will first briefly review the theory of three-dimensional multipole field expansion in a straight, circular cylinder geometry. Our discussion is limited to three-dimensional multipole expansion of the magnetic scalar potential, rather than the vector potential. After a brief review of theory, we will then present detailed expansion techniques with the example in Figure 3.2. The linear and high-order terms in the expansion are discussed.

4.1. Review of Theory

It is well known that in a current-free region the magnetic scalar potential Φ_m satisfies the Laplace's equation

$$\nabla^2 \Phi_m = 0 . \tag{4.1}$$

In cylindrical coordinates (r, θ, z) the solution of Eq. (4.1) can be expanded in terms of the eigenfunctions of the operator $\partial^2 / \partial\theta^2$ as

$$\Phi_m = \sum_{m=0}^{\infty} \phi_{m,s}(r,z) Sin(m\theta) + \phi_{m,c}(r,z) Cos(m\theta), \tag{4.2a}$$

where

$$\phi_{m,s}(r,z) = \frac{1}{\sqrt{2\pi}} \int_{-\infty}^{\infty} dk \exp(ikz) I_m(kr) b_m(k), \tag{4.2b}$$

$$\phi_{m,c}(r,z) = \frac{1}{\sqrt{2\pi}} \int_{-\infty}^{\infty} dk \exp(ikz) I_m(kr) a_m(k). \tag{4.2c}$$

Equation (4.2a) is usually referred to as the three-dimensional multipole expansion for the magnetic scalar potential. The 'sine-like' and 'cosine-like' terms are the normal and skew components. The integer m is the order of the multipoles. For example, m = 0 corresponds to a pure solenoid, m = 1 to a dipole, and m = 2 to a quadrupole, etc. The solenoidal field is described by the 'cosine-like' term only, i.e., $\phi_{0,c}(r, z)$. The physical meaning of the functions $b_m(k)$ and $a_m(k)$ in Eqs. (4.2b, c) will become clear later. I_m is the modified Bessel function of the first kind of order m, which can be expressed by a Taylor expansion

$$I_m(x) = \sum_{\ell=0}^{\infty} \frac{1}{\ell!(m+\ell)!} \left(\frac{x}{2}\right)^{2\ell+m} . \tag{4.3}$$

With the help of Eq. (4.3), we can rewrite Eqs. (4.2b, 4.2c) as

$$\phi_{m,\alpha}(r,z)=\sum_{\ell=0}^{\infty}(-1)^{\ell}\frac{m!}{2^{2\ell}\,\ell!(\ell+m)!}C_{m,\alpha}^{[2\ell]}(z)r^{2\ell+m}, \tag{4.4}$$

where $\alpha = s$ or c, and the functions $C_{m,\alpha}^{[2\ell]}(z)$ are defined as

$$C_{m,s}^{[2\ell]}(z)=\frac{(-1)^{\ell}}{2^m\,m!}\frac{1}{\sqrt{2\pi}}\int_{-\infty}^{\infty}dk\,\exp(ikz)k^{2\ell+m}b_m(k), \tag{4.5a}$$

$$C_{m,c}^{[2\ell]}(z)=\frac{(-1)^{\ell}}{2^m\,m!}\frac{1}{\sqrt{2\pi}}\int_{-\infty}^{\infty}dk\,\exp(ikz)k^{2\ell+m}a_m(k). \tag{4.5b}$$

It is easy to verify that

$$C_{m,\alpha}^{[2\ell+2]}(z)=\frac{d^2}{dz^2}C_{m,\alpha}^{[2\ell]}(z).$$

Therefore, if $C_{m,\alpha}^{[0]}(z)\equiv C_{m,\alpha}(z)$ is known, all the expansion coefficients in Eq. (4.4) can be found by successive differentiation. The functions $C_{m,\alpha}(z)$ are called the generalized gradients. Their derivatives constitute the so-called pseudo-multipoles or pseudo-harmonics.

The magnetic field can be found by the gradients of the magnetic scalar potential: $\mathbf{B}=\nabla\Phi_m$. The convention here is to use a + sign. The field components in cylindrical coordinates are

$$\begin{aligned}B_r &= \sum_{m=0}\sum_{\ell=0}^{\infty}(-1)^{\ell}\frac{m!(2\ell+m)}{2^{2\ell}\,\ell!(\ell+m)!}r^{2\ell+m-1}\times\\ &\times\left\{C_{m,s}^{[2\ell]}(z)Sin(m\theta)+C_{m,c}^{[2\ell]}(z)Cos(m\theta)\right\},\end{aligned} \tag{4.6a}$$

$$\begin{aligned}B_\theta &= \sum_{m=0}\sum_{\ell=0}^{\infty}(-1)^{\ell}\frac{m!m}{2^{2\ell}\,\ell!(\ell+m)!}r^{2\ell+m-1}\times\\ &\times\left\{C_{m,s}^{[2\ell]}(z)Cos(m\theta)-C_{m,c}^{[2\ell]}(z)Sin(m\theta)\right\},\end{aligned} \tag{4.6b}$$

$$B_z = \sum_{m=0}^{\infty} \sum_{\ell=0} (-1)^{\ell} \frac{m!}{2^{2\ell}\, \ell!(\ell+m)!} r^{2\ell+m} \times$$
$$\times \left\{ C_{m,s}^{[2\ell+1]}(z) Sin(m\theta) + C_{m,c}^{[2\ell+1]}(z) Cos(m\theta) \right\} . \quad (4.6c)$$

Note that the solenoidal field (m = 0) does not contribute to the azimuthal field B_θ, though we keep the formulas in the same format. Recall that in a two-dimensional multipole expansion described in Section 2.2, the radial dependence of the transverse field harmonics of order m is r^{m-1}. However, in a three-dimensional multipole expansion this applies only to the terms associated with the generalized gradients $C_{m,\alpha}(z)$ ($\ell = 0$). There are still many more terms of the pseudo-harmonics that vary radially as $r^{2\ell+m-1}$ (ℓ is not equal to 0). For example, in a quadrupole (m = 2) the component $C_{2,\alpha}^{[2]} r^3$ ($\ell = 1$) is called a pseudo-octupole since its azimuthal variation follows a quadrupole field while its radial dependence resembles an octupole. The pseudo-harmonics are produced by the fringe fields. If the field is uniform in the z-direction, then the derivatives of the generalized gradients vanish and the pseudo-harmonics disappear, and therefore we recover the two-dimensional multipoles.

According to the Uniqueness theorem, the magnetic field in a volume of a current-free region is uniquely determined by the boundary conditions. If the magnetic field is known on the surface of a very long cylinder co-axial with the magnet z-axis, we should be able to find the field at any point within the cylinder by the method of three-dimensional multipole expansion. Suppose that the radial field B_r on a cylindrical surface of radius R is found from simulation or experiment, we first analyze the field via the Fourier expansion to obtain

$$B_r(R,\theta,z) = \sum_{m=0}^{\infty} \mathcal{B}_m(R,z) Sin(m\theta) + \mathcal{A}_m(R,z) Cos(m\theta), \quad (4.7)$$

where $\mathcal{B}_m(R, z)$ and $\mathcal{A}_m(R, z)$ are the amplitudes of the normal and skew components of $B_r(R,\theta, z)$ harmonics of order m. In comparison with B_r obtained from Eq. (4.2a) and Eq. (4.7), we find

$$b_m(k) = \frac{1}{\sqrt{2\pi}} \int_{-\infty}^{\infty} dz \exp(-ikz) \frac{\mathcal{B}_m(R,z)}{kI_m'(kR)}, \quad (4.8a)$$

$$a_m(k) = \frac{1}{\sqrt{2\pi}} \int_{-\infty}^{\infty} dz \exp(-ikz) \frac{\mathcal{A}_m(R,z)}{kI_m'(kR)}. \tag{4.8b}$$

These equations show that the functions $b_m(\mathrm{k})$ and $a_m(\mathrm{k})$ in the multipole expansion, Eqs. (4.2b and 4.2c), are the Fourier transforms of $\mathcal{B}_m(R, z)$ and $\mathcal{A}_m(R, z)$ weighted by a function $1/\mathrm{k}I_m'(kR)$. Inserting Eqs. (4.8a, b) into Eqs. (4.5a, b) yields the generalized gradients

$$C_{m,s}(z) = \frac{1}{2^m m!} \frac{1}{\sqrt{2\pi}} \int_{-\infty}^{\infty} dk \exp(ikz) \frac{k^{m-1}}{I_m'(kR)} \widetilde{\mathcal{B}}_m(R,k), \tag{4.9a}$$

$$C_{m,c}(z) = \frac{1}{2^m m!} \frac{1}{\sqrt{2\pi}} \int_{-\infty}^{\infty} dk \exp(ikz) \frac{k^{m-1}}{I_m'(kR)} \widetilde{\mathcal{A}}_m(R,k). \tag{4.9b}$$

Here $\widetilde{\mathcal{B}}_m(R,k)$ and $\widetilde{\mathcal{A}}_m(R,k)$ are the Fourier transforms of $\mathcal{B}_m(R, z)$ and $\mathcal{A}_m(R, z)$ as given by

$$\widetilde{\mathcal{B}}_m(R,k) = \frac{1}{\sqrt{2\pi}} \int_{-\infty}^{\infty} dz \exp(-ikz) \mathcal{B}_m(R,z), \tag{4.10a}$$

$$\widetilde{\mathcal{A}}_m(R,k) = \frac{1}{\sqrt{2\pi}} \int_{-\infty}^{\infty} dz \exp(-ikz) \mathcal{A}_m(R,z). \tag{4.10b}$$

In summary, a three-dimensional multipole expansion of magnetic fields starts with the field data on a cylindrical surface, e.g. $B_r(R, z)$ from simulation in our discussion. The surface field data from measurements can equally be employed for the same purpose. We then decompose $B_r(R, z)$ into its Fourier components $\mathcal{B}_m(R, z)$ and $\mathcal{A}_m(R, z)$, as indicated in Eq. (4.7). The next step is to obtain $\widetilde{\mathcal{B}}_m(R,k)$ and $\widetilde{\mathcal{A}}_m(R,k)$ by the Fourier transformation according to (4.10a) and (4.10b). The inverse Fourier transforms of $\widetilde{\mathcal{B}}_m(R,k)$ and $\widetilde{\mathcal{A}}_m(R,k)$, weighted by a function $k^{m-1}/I_m'(kR)$, yield the generalized gradients $C_{m,\alpha}(z)$ according to Eqs. (4.9a, b). Finally, the magnetic field components can be obtained by Eqs. (4.6a, b, c). Although we use the radial magnetic component $B_r(R, z)$ to illustrate the applications of the theory, the azimuthal component $B_\theta(R, z)$ can also be employed for the same task.

4.2. EXPANSION TECHNIQUES

Taking the simulation model of the SNS injection chicane dipoles D2 and D3 presented in Section 3.3 as an example, we first need to calculate the surface fields on a long cylinder by using OPERA-3D post processor in the simulation model. There are two basic field calculation methods in OPERA-3D post processor: nodal interpolation and integration. The former, which is the default setting and is more commonly used, offers fast computation but less accuracy. The latter is slow but should give an order of magnitude improvement in accuracy, since it evaluates the iron magnetizations from the solution and then computes the fields by integration of $\nabla \times (\mathbf{M} \times \nabla(1/r))$ [13]. The method of calculating fields by integration is employed in this practice.

In the three-dimensional multipole expansion of the D2 and D3 fields, we construct a long cylinder of radius $R = 10$ cm, which is coaxial with the magnetic axis. We then calculate the radial field $B_r(R, z)$ on the surface of the cylinder, starting from z = -240 cm and ending at z = 420 cm in a step size of 0.25 cm. There are a total of 2641 data points in the z direction. At each z-position, $B_r(R, z)$ is computed along a circle of radius $R = 10$ cm at a step of 0.5 degree, which yields 720 data points for the Fourier decomposition. The $B_r(R, z)$ harmonic components up to m = 10 (the 20th pole) are computed.

Figure 4.1 shows the dominant term $\mathcal{B}_1$ versus z, which is the normal dipole term for m = 1. To some extent, $\mathcal{B}_1$ resembles the dipole field component B_y on the z-axis as shown in Figure 3.3. However, $\mathcal{B}_1$ contains highly nonlinear terms, i.e. the pseudo-harmonics, because it is evaluated at $R = 10$ cm. The other two important components of the $B_r(R, z)$ harmonics are the skew term $\mathcal{A}_1$ for m = 1 and the solenoidal term $\mathcal{A}_0$ as plotted in Figures 4.2 and 4.3. The former resembles B_x on the z-axis while the latter has connection with B_z on the z-axis, which will be made clearer in the next section.

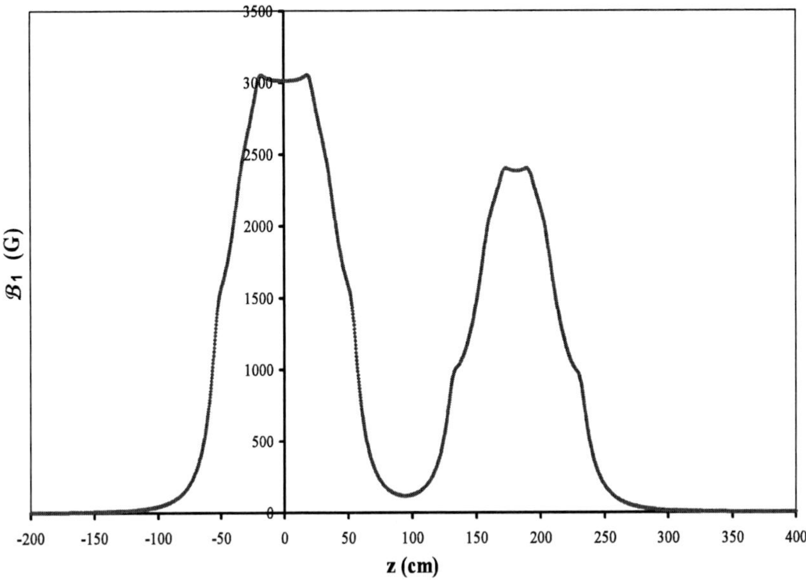

Figure 4.1. Dominant component $\mathcal{B}_1$ in the surface field B_r.

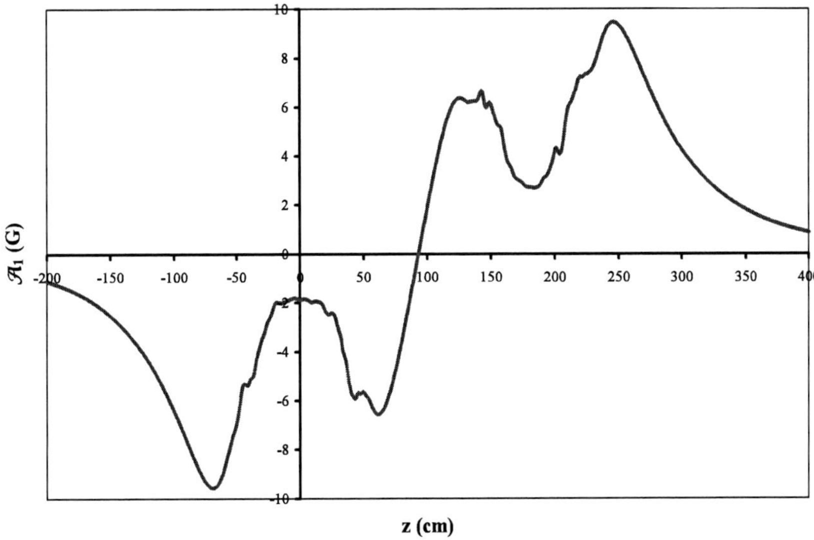

Figure 4.2. Component $\mathcal{A}_1$ in the surface field B_r.

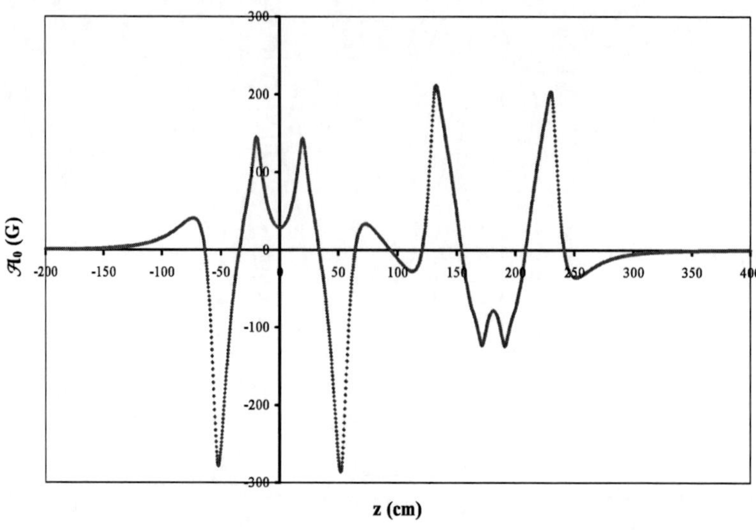

Figure 4.3. Component $\mathcal{A}_0$ in the surface field B_r.

The next step is to perform the Fourier transforms of $\mathcal{B}_m(R, z)$ and $\mathcal{A}_m(R, z)$ to yield $\widetilde{\mathcal{B}}_{m,\alpha}(R,k)$ and $\widetilde{\mathcal{A}}_{m,\alpha}(R,k)$ as described in Eq. (4.10). Due to their complex nature, the Fourier transforms produce two terms for each harmonic component of $B_r(R, \theta, z)$: one associated with the sine term and the other with the cosine term. The wave number k is chosen from 0 to 2 (1/cm) in 4000 steps. The transforms yield spectrum-like distributions of the fields. The dominant terms $\widetilde{\mathcal{B}}_{1,s}(R,k)$ and $\widetilde{\mathcal{B}}_{1,c}(R,k)$ are plotted in Figure 4.4 and the other two components $\widetilde{\mathcal{A}}_{1,\alpha}(R,k)$ and $\widetilde{\mathcal{A}}_{0,\alpha}(R,k)$ are shown in Figures 4.5 and 4.6. Having done this, the generalized gradients can be computed by using the inverse Fourier transforms of $\widetilde{\mathcal{B}}_{m,\alpha}(R,k)$ and $\widetilde{\mathcal{A}}_{m,\alpha}(R,k)$, weighted by a function $k^{m-1}/I_m'(kR)$ as expressed in Eqs. (4.9a, b). Note that Eqs. (4.9a, b) have a singularity at $k = 0$ for m = 0, which is the case of a solenoidal field. There are three options to avoid this problem. First, we can make k sufficiently close but not equal to zero. The magnitude of the resulting $C_0(z) = C_{0,c}(z)$ depends on the minimum k chosen, but its first derivative $C_0'(z)$ remains approximately the same. It is $C_0'(z)$ rather than $C_0(z)$ that has a physical meaning. Second, we can calculate $C_0''(z)$ in the inverse Fourier transform and then to get $C_0'(z)$ via numerical integration. Some errors will still occur in numerical integration. The third option is an easy method in practice and will be described in Section 4.3.

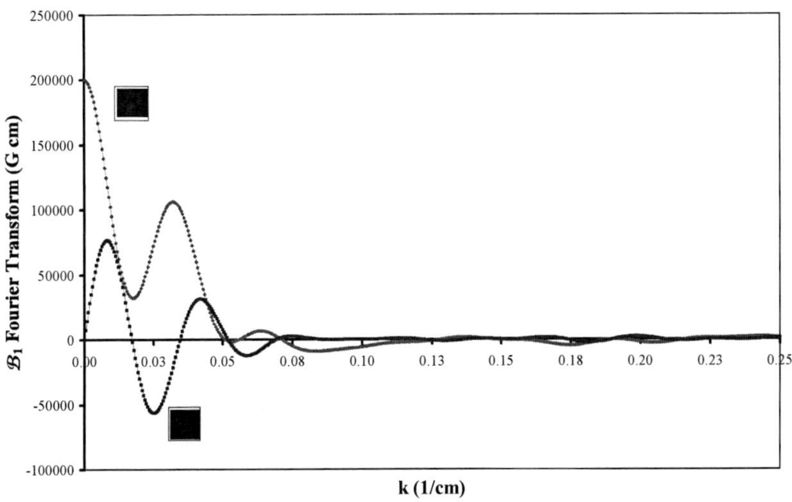

Figure 4.4. $\widetilde{\mathcal{B}}_{1,\alpha}(R,k)$ versus k.

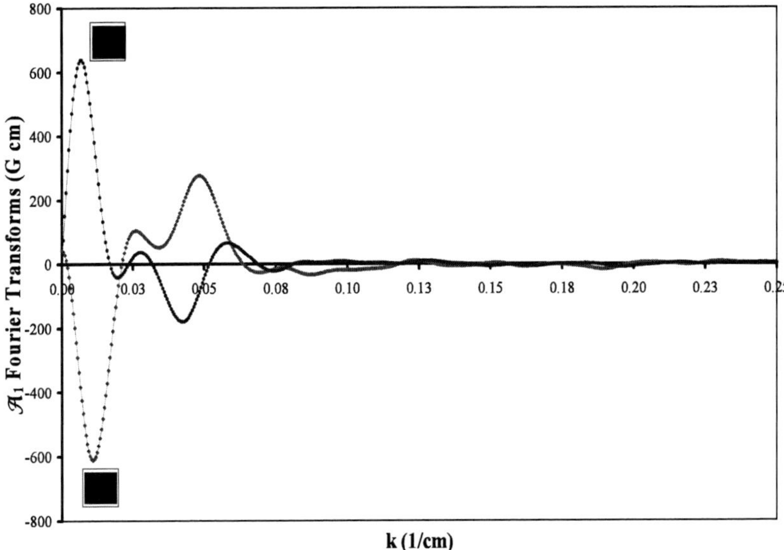

Figure 4.5. $\widetilde{\mathcal{A}}_{1,\alpha}(R,k)$ versus k.

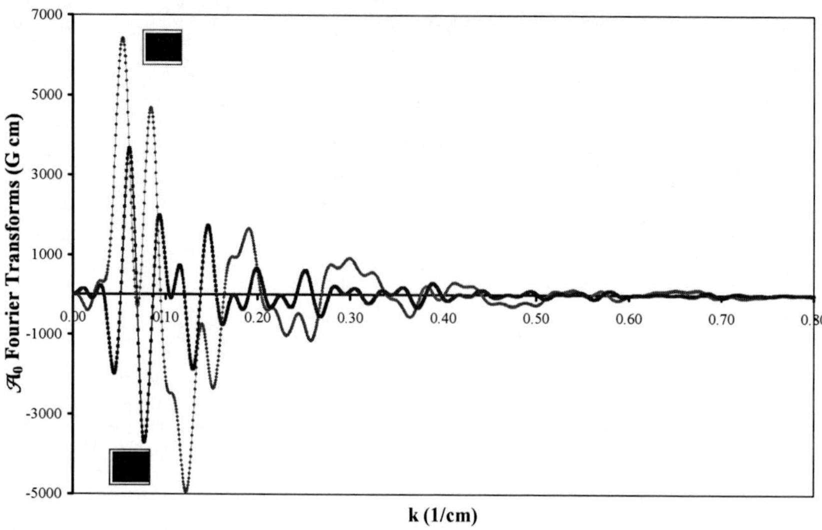

Figure 4.6. $\widetilde{\mathcal{A}}_{0,\alpha}(R,k)$ versus k.

With the generalized gradients calculated above, the remaining work is to perform successive differentiation to obtain the desired pseudo-harmonics. This is done via a Mathematica program [14], which first fits the numerical data of $C_{m,\alpha}(z)$ into interpolation functions of a certain order. Although numerical differentiation in Mathematica is straight forward, numerical errors are always introduced and numerical oscillations from higher-order differentiations are often large enough to invalidate the results. A way to solve this problem is to calculate the even derivatives $C_{m,\alpha}^{[2\ell]}(z)$ according to

$$C_{m,s}^{[2\ell]}(z) = \frac{1}{2^m m!} \frac{1}{\sqrt{2\pi}} \int_{-\infty}^{\infty} dk \exp(ikz) \frac{(-1)^{2\ell} k^{m+2\ell-1}}{I_m'(kR)} \widetilde{\mathcal{B}}_m(R,k), \tag{4.11a}$$

$$C_{m,c}^{[2\ell]}(z) = \frac{1}{2^m m!} \frac{1}{\sqrt{2\pi}} \int_{-\infty}^{\infty} dk \exp(ikz) \frac{(-1)^{2\ell} k^{m+2\ell-1}}{I_m'(kR)} \widetilde{\mathcal{A}}_m(R,k). \tag{4.11b}$$

Similarly, we can obtain the odd derivatives $C_{m,\alpha}^{[2\ell+1]}(z)$ by

$$C_{m,s}^{[2\ell+1]}(z) = \frac{1}{2^m m!} \frac{1}{\sqrt{2\pi}} \int_{-\infty}^{\infty} dk \exp(ikz) \frac{i(-1)^{2\ell} k^{m+2\ell}}{I_m'(kR)} \tilde{\mathcal{B}}_m(R,k), \tag{4.11c}$$

$$C_{m,c}^{[2\ell+1]}(z) = \frac{1}{2^m m!} \frac{1}{\sqrt{2\pi}} \int_{-\infty}^{\infty} dk \exp(ikz) \frac{i(-1)^{2\ell} k^{m+2\ell}}{I_m'(kR)} \tilde{\mathcal{A}}_m(R,k). \tag{4.11d}$$

This method is very effective in eliminating numerical oscillations and has been proven to work well for all the trials up to $2\ell = 10$ in this practice. The second option to calculate $C_0'(z)$ in the previous paragraph follows this technique.

4.3. On-Axis Gradients

The generalized gradients $C_0'(z)$, $C_{1,s}(z)$, and $C_{1,c}(z)$ have special physical meanings. They express the on-axis magnetic field components $B_z(z)$, $B_y(z)$, and $B_x(z)$ at $x = y = 0$, respectively. Since these gradients are obtained via a number of mathematical manipulations, it would be interesting to know if they agree with the on-axis field data computed directly from the OPERA-3D simulation. This is a way to verify the theory and check all the procedures of mathematical calculations. In Figure 4.7(a) we compare the generalized gradient $C_{1,s}(z)$ with the field data $B_y(z)$ at $x = y = 0$ obtained directly from OPERA-3D. The agreement looks remarkably good. In order to see the discrepancies of the two sets of data more clearly, we plot their differences in Figure 4.7(b). The maximum difference is about 0.017 G out of a total field range of 3009.36 G. This corresponds to a relative error at 10^{-6} level, which is attributed to numerical errors. Indeed, the generalized gradient $C_{1,s}(z)$ is an accurate representation of the dipole field $B_y(z)$ on the z-axis. Figure 4.8 compares the generalized gradient $C_0'(z)$ and the OPERA-3D data $B_z(z)$ at $x = y = 0$, and Figure 4.9 compares the generalized gradient $C_{1,c}(z)$ and the OPERA-3D data $B_x(z)$ at $x = y = 0$. Again, the agreements are excellent. In principle, the three-dimensional multipole expansion is insensitive to the raw data errors produced in simulations or experiments. This is due to a basic property of the solutions of Laplace's equation: the value of the magnetic scalar potential Φ_m at some interior point is an appropriately weighted average of its values over any surrounding boundary.

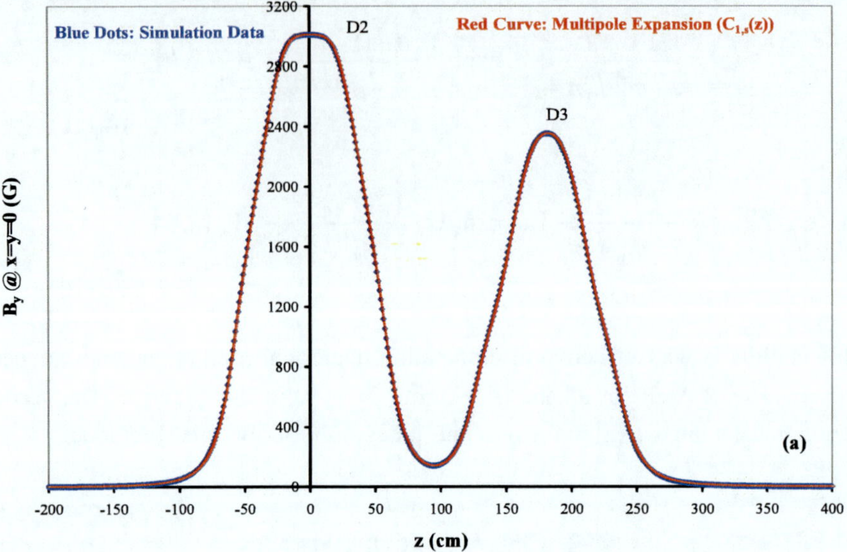

Figure 4.7. (a) B_y at x = y = 0 versus z.

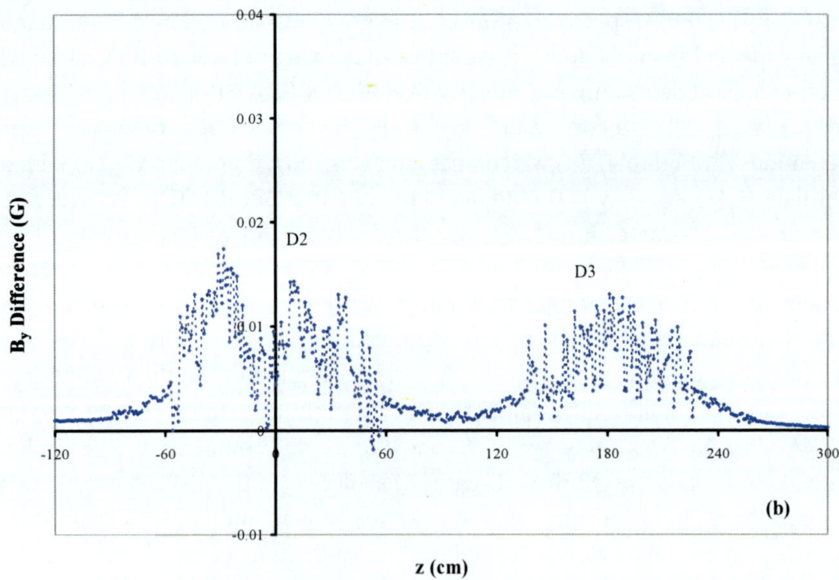

Figure 4.7. (b) Difference in B_y between multipole expansion and simulation data.

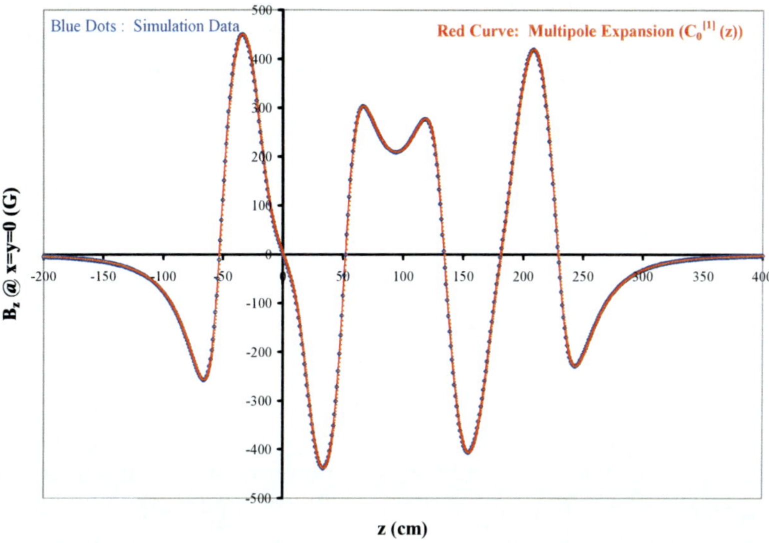

Figure 4.8. B_z at x = y = 0 versus z.

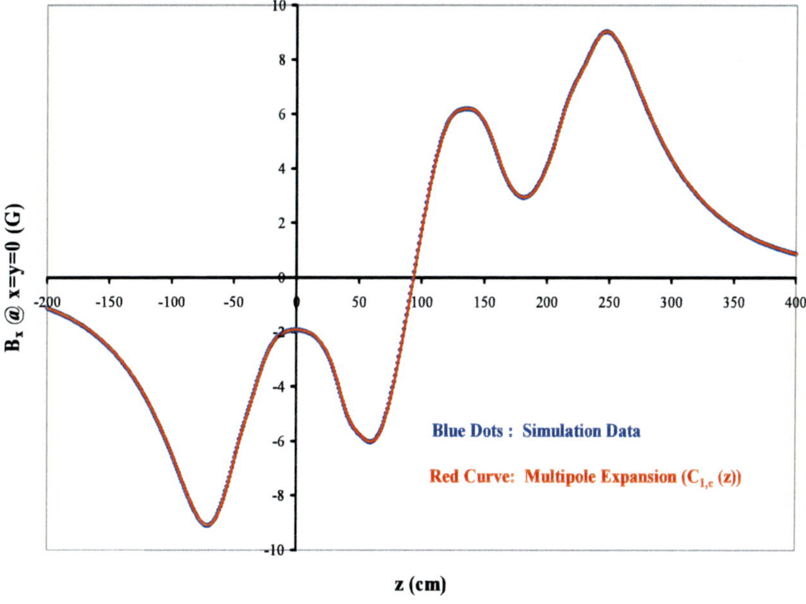

Figure 4.9. B_x at x = y = 0 versus z.

The above results have shown that the generalized gradients $C_0'(z)$, $C_{1,s}(z)$, and $C_{1,c}(z)$ indeed represent the on-axis magnetic field components $B_z(z)$, $B_y(z)$, and $B_x(z)$ at x = y = 0, respectively. Thus, we can in fact obtain these generalized gradients directly from the three-dimensional simulation data rather than from using complicated expansion procedures. This method is easier, faster, and more accurate. This is especially true for $C_0'(z)$ since the other two options involve more numerical errors.

4.4. A 5TH-ORDER REPRESENTATION

To calculate magnetic fields at any point off the z-axis within the cylinder, Eqs. (4.6a, b, c) have to be used. In practice, a cut to a certain order of the azimuthal harmonic number m and the radial power number l has to be made. A 5th-order representation of magnetic fields covers the multipoles up to the regular dodecapoles and pseudo-dodecapoles. In this case, a total of 13 generalized gradients are required. They include the solenoidal component and both the normal and skew terms of the dipole, quadrupole, sextupole, octupole, decapole, and dodecapole components. These generalized gradients can be stored in computer as interpolation functions. Their derivatives can be easily calculated mathematically to yield the pseudo-multipoles for subsequent calculations. Thus, much less memory is required and the field calculations become quasi-analytical. More importantly, this representation of magnetic fields can lead to desired particle optics up to the 5th-order. The explicit expressions for this representation, resulted from Eqs. (4.6a,b,c) in cylindrical coordinates, are given below

$$
\begin{aligned}
B_r = {} & [-\frac{1}{2}C_0^{[2]}r + \frac{1}{16}C_0^{[4]}r^3 - \frac{1}{384}C_0^{[6]}r^5 + ...] + \\
& + [C_{1,s} - \frac{3}{8}C_{1,s}^{[2]}r^2 + \frac{5}{192}C_{1,s}^{[4]}r^4 - ......]Sin(\theta) + [C_{1,c} - \frac{3}{8}C_{1,c}^{[2]}r^2 + \frac{5}{192}C_{1,c}^{[4]}r^4 - ......]Cos(\theta) + \\
& + [2C_{2,s}r - \frac{1}{3}C_{2,s}^{[2]}r^3 + \frac{1}{64}C_{2,s}^{[4]}r^5 - ......]Sin(2\theta) + [2C_{2,c}r - \frac{1}{3}C_{2,c}^{[2]}r^3 + \frac{1}{64}C_{2,c}^{[4]}r^5 - ......]Cos(2\theta) + \\
& + [3C_{3,s}r^2 - \frac{5}{16}C_{3,s}^{[2]}r^4 + ...]Sin(3\theta) + [3C_{3,c}r^2 - \frac{5}{16}C_{3,c}^{[2]}r^4 + ...]Cos(3\theta) + \\
& + [4C_{4,s}r^3 - \frac{3}{10}C_{4,s}^{[2]}r^5 + ...]Sin(4\theta) + [4C_{4,c}r^3 - \frac{3}{10}C_{4,c}^{[2]}r^5 + ...]Cos(4\theta) + \\
& + [5C_{5,s}r^4 - ......]Sin(5\theta) + [5C_{5,c}r^4 - ......]Cos(5\theta) + \\
& + [6C_{6,s}r^5 - ......]Sin(6\theta) + [6C_{6,c}r^5 - ......]Cos(6\theta) + \\
& + ...
\end{aligned}
\tag{4.12a}
$$

$$
\begin{aligned}
B_\theta = {} & [C_{1,s} - \frac{1}{8}C_{1,s}^{[2]}r^2 + \frac{1}{192}C_{1,s}^{[4]}r^4 - ......]Cos(\theta) - [C_{1,c} - \frac{1}{8}C_{1,c}^{[2]}r^2 + \frac{1}{192}C_{1,c}^{[4]}r^4 - ......]Sin(\theta) + \\
& + [2C_{2,s}r - \frac{1}{6}C_{2,s}^{[2]}r^3 + \frac{1}{192}C_{2,s}^{[4]}r^5 - ......]Cos(2\theta) - [2C_{2,c}r - \frac{1}{6}C_{2,c}^{[2]}r^3 + \frac{1}{192}C_{2,c}^{[4]}r^5 - ......]Sin(2\theta) + \\
& + [3C_{3,s}r^2 - \frac{3}{16}C_{3,s}^{[2]}r^4 + ...]Cos(3\theta) - [3C_{3,c}r^2 - \frac{3}{16}C_{3,c}^{[2]}r^4 + ...]Sin(3\theta) + \\
& + [4C_{4,s}r^3 - \frac{1}{5}C_{4,s}^{[2]}r^5 + ...]Cos(4\theta) - [4C_{4,c}r^3 - \frac{1}{5}C_{4,c}^{[2]}r^5 + ...]Sin(4\theta) + \\
& + [5C_{5,s}r^4 - ......]Cos(5\theta) - [5C_{5,c}r^4 - ......]Sin(5\theta) + \\
& + [6C_{6,s}r^5 - ......]Cos(6\theta) - [6C_{6,c}r^5 - ......]Sin(6\theta) + \\
& + ...
\end{aligned}
\tag{4.12b}
$$

$$
\begin{aligned}
B_z = {} & [C_0^{[1]} - \frac{1}{4}C_0^{[3]}r^2 + \frac{1}{64}C_0^{[5]}r^4 - ......] + \\
& + [C_{1,s}^{[1]}r - \frac{1}{8}C_{1,s}^{[3]}r^3 + \frac{1}{192}C_{1,s}^{[5]}r^5 - ......]Sin(\theta) + [C_{1,c}^{[1]}r - \frac{1}{8}C_{1,c}^{[3]}r^3 + \frac{1}{192}C_{1,c}^{[5]}r^5 - ......]Cos(\theta) + \\
& + [C_{2,s}^{[1]}r^2 - \frac{1}{12}C_{2,s}^{[3]}r^4 + ...]Sin(2\theta) + [C_{2,c}^{[1]}r^2 - \frac{1}{12}C_{2,c}^{[3]}r^4 + ...]Cos(2\theta) + \\
& + [C_{3,s}^{[1]}r^3 - \frac{1}{16}C_{3,s}^{[3]}r^5 + ...]Sin(3\theta) + [C_{3,c}^{[1]}r^3 - \frac{1}{16}C_{3,c}^{[3]}r^5 + ...]Cos(3\theta) + \\
& + [C_{4,s}^{[1]}r^4 - ......]Sin(4\theta) + [C_{4,c}^{[1]}r^4 - ......]Cos(4\theta) + \\
& + [C_{5,s}^{[1]}r^5 - ......]Sin(5\theta) + [C_{5,c}^{[1]}r^5 - ......]Cos(5\theta) + \\
& + ...
\end{aligned}
\tag{4.12c}
$$

We verify these equations by comparing their results with the field data obtained directly from the OPERA-3D simulation. A transformation of Eqs. (4.12a,b,c) to Cartesian coordinates is made first for numerical calculations. Figures 4.10 and 4.11 plot the field components B_y versus y at x = z = 0, and B_x versus x at y = z = 0, respectively. The agreement between the simulation data and the 5th-order representation is very good. In Figures 4.12 and 4.13 we plot B_y versus x at y = z = 0, and B_x versus y at x = z = 0. Again, we see good agreement in general between the 5th-order representation and the simulation data, especially in the near-axis region. The deviation of the 5th-order representation results from the direct simulation data becomes large far away from the magnet axis. This is due to the cut off higher-order terms in Eqs. (4.12a,b,c). A 9th-order (up to 20th pole) approximation can yield much better agreement, as shown by the green curve in Figures 4.12 and 4.13. This higher-order effect will be explained in more detail in the next section.

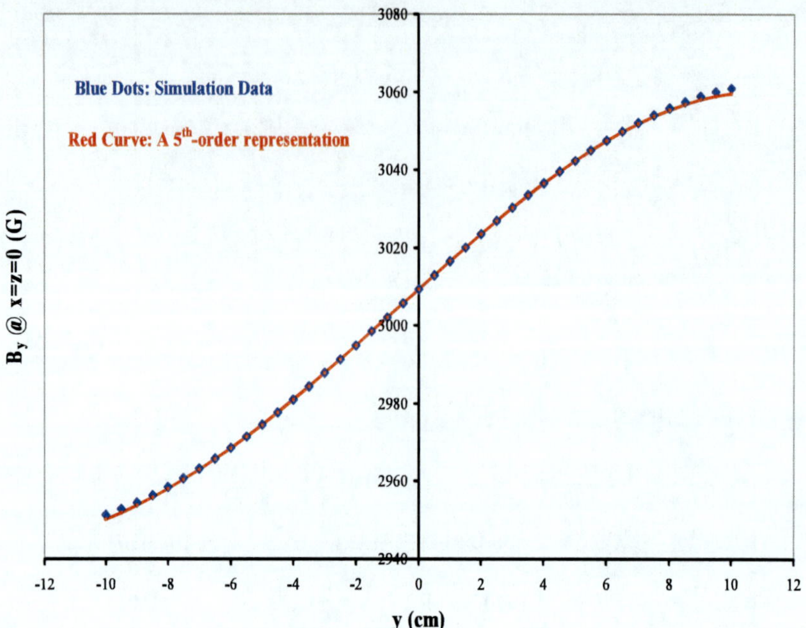

Figure 4.10. B_y versus y at x = z = 0.

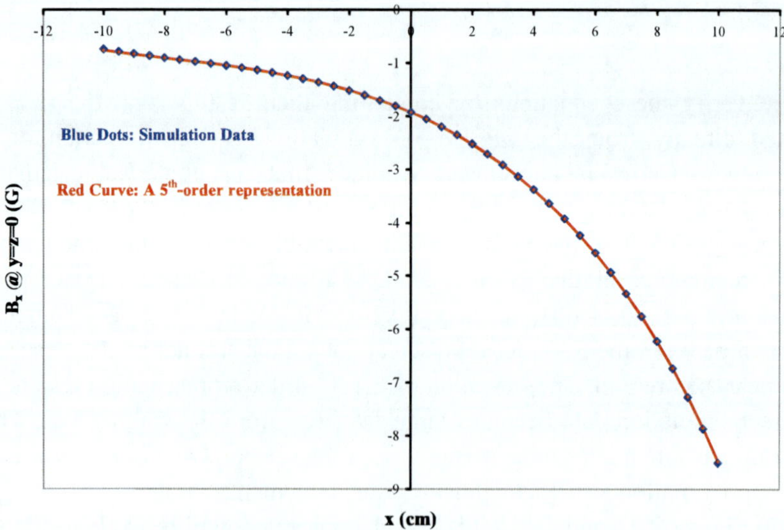

Figure 4.11. B_x versus x at y = z = 0.

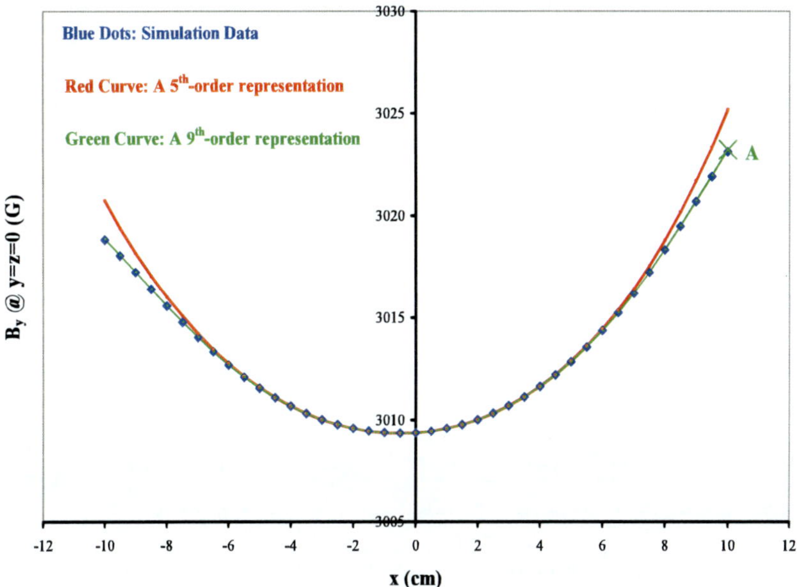

Figure 4.12. B_y versus x at y = z = 0.

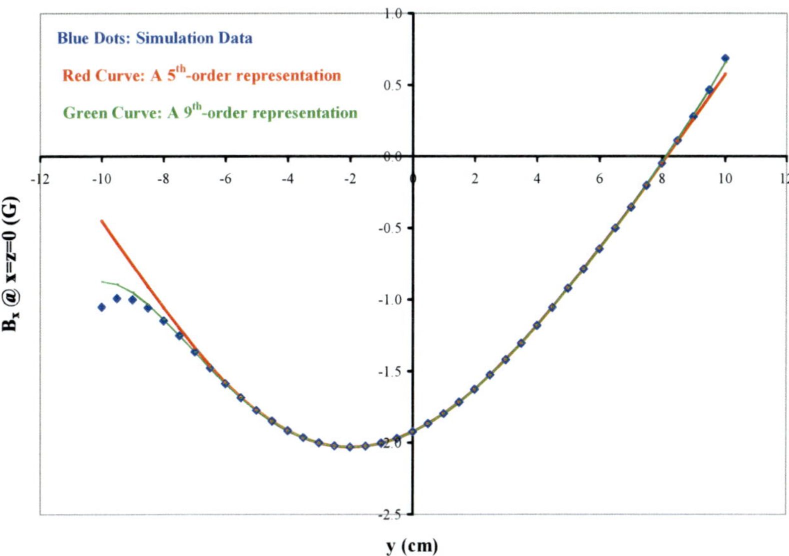

Figure 4.13. B_x versus y at x = z = 0.

4.5. Higher-Order Effects

Figures 4.12 and 4.13 reveal that the numerical results calculated from the 5th-order representation, i.e. from Eqs. (4.12a,b,c), deviate more from the direct simulation field data when space points are far away from the magnet axis. This is caused by cutting off higher-order terms (m > 6 and beyond r^5) and can be corrected. In order to demonstrate the effect of the higher-order terms more clearly, we write down explicitly a 9th-order formula for B_y at y = z = 0 versus x when x is on the positive axis:

$$
\begin{aligned}
B_y \approx\; & [C_{1,s}(0) - \frac{1}{8}C_{1,s}^{[2]}(0)x^2 + \frac{1}{192}C_{1,s}^{[4]}(0)x^4 - \frac{1}{9216}C_{1,s}^{[6]}(0)x^6 + \frac{1}{737280}C_{1,s}^{[8]}(0)x^8] + \\
& + [2C_{2,s}(0)x - \frac{1}{6}C_{2,s}^{[2]}(0)x^3 + \frac{1}{192}C_{2,s}^{[4]}(0)x^5 - \frac{1}{11520}C_{2,s}^{[6]}(0)x^7 + \frac{1}{1105920}C_{2,s}^{[8]}(0)x^9] + \\
& + [3C_{3,s}(0)x^2 - \frac{3}{16}C_{3,s}^{[2]}(0)x^4 + \frac{3}{640}C_{3,s}^{[4]}(0)x^6 - \frac{1}{15360}C_{3,s}^{[6]}(0)x^8] + \\
& + [4C_{4,s}(0)x^3 - \frac{1}{5}C_{4,s}^{[2]}(0)x^5 + \frac{1}{240}C_{4,s}^{[4]}(0)x^7 - \frac{1}{20160}C_{4,s}^{[6]}(0)x^9] + \\
& + [5C_{5,s}(0)x^4 - \frac{5}{24}C_{5,s}^{[2]}(0)x^6 + \frac{5}{1344}C_{5,s}^{[4]}(0)x^8] + \\
& + [6C_{6,s}(0)x^5 - \frac{3}{14}C_{6,s}^{[2]}(0)x^7 + \frac{3}{896}C_{6,s}^{[4]}(0)x^9] + \\
& + [7C_{7,s}(0)x^6 - \frac{7}{32}C_{7,s}^{[2]}(0)x^8] + \\
& + [8C_{8,s}(0)x^7 - \frac{4}{9}C_{8,s}^{[2]}(0)x^9] + \\
& + 9C_{9,s}(0)x^8 + 10C_{10,s}(0)x^9
\end{aligned}
\tag{4.13}
$$

The green curve for x > 0 in Figure 4.12 is generated by Eq. (4.13), which agrees with the simulation data very well. It would be interesting to know the contributions to B_y from each azimuthal harmonic term up to m = 10 (the 20th pole) and from each radial power up to x^9. This is listed in Table 4.1 for point A, which is indicated by a cross at x = 10 cm in Figure 4.12. The summations on the right-most column give the contributions from different orders in r, while the summations on the bottom rows give the contributions from different harmonic number m for three different orders in *r*. A full 9th-order representation yields B_y = 3023.23 G at x = 10 cm and y = z = 0. As compared to the simulation data of 3023.13 G at the same point, the difference is only 30 ppm. Therefore, we have demonstrated that we can approximate the simulation data to very high accuracy by higher-order representations in the expansion.

Table 4.1. Contributions of multipole terms to B_y @ x = 10 cm, y = z = 0

B_y @ x = 10 cm, y = z = 0

Multipole expansion:

	m = 1	m = 2	m = 3	m = 4	m =5	m =6	m =7	m =8	m =9	m =10	Sum:
r^0	3009.360										3009.360
r^1		1.049									1.049
r^2	1.792		8.950								10.742
r^3		0.015		0.868							0.883
r^4	-0.466		0.700		2.619						2.853
r^5		0.004		-0.012		0.302					0.294
r^6	0.056		-0.106		0.056		-1.250				-1.243
r^7		0.003		-0.009		0.023		-0.106			-0.089
r^8	-0.002		0.004		-0.001		-0.001		-0.664		-0.665
r^9		0.001		-0.005		0.013		-0.011		0.042	0.041
Sum:											
5th	3010.687	1.067	9.650	0.857	2.619	0.302	0.000	0.000	0.000	0.000	3025.18
7th	3010.743	1.070	9.544	0.848	2.675	0.325	-1.250	-0.106	0.000	0.000	3023.85
9th	3010.741	1.071	9.548	0.842	2.674	0.338	-1.251	-0.117	-0.664	0.042	3023.23

Simulation data: B_y = 3023.13 G

In summary, the results from the three-dimensional multipole expansion are very satisfactory and can achieve the level of accuracy desired. In general, a 5th-order approximation can make a fairly accurate analytical representation of the real field distribution. The accuracy can be improved even more for the field points far away from the axis by adding higher-order terms of m>6. The three-dimensional multipole expansion has given much insight into the magnet physics in terms of harmonics and pseudo-harmonics. The three-dimensional multipole expansion is analytical in nature when we fit the generalized gradients of numerical data into interpolation functions. This feature will prove to be very useful in further beam dynamics study. The theory and techniques presented in this chapter can be equally applied to experimental data on a cylindrical surface, as well as to other accelerator magnets, such as quadrupoles, sextupoles, octupoles, and etc.

REFERENCES

[1] P. M. Morse and H. Feshbach, *Methods of Theoretical Physics*, McGraw-Hill, New York, 1953.

[2] J. Cobb and R. Cole, "Spectroscopy of quadrupole magnets," in Proceedings of the International Symposium on Magnet Technology, P. 431, Stanford, CA, 1965.

[3] J. K. Cobb, A. W. Burfine and D. R. Jensen, "Determination of the quality of multipole magnets," in Proceedings of the 2nd International Conference on Magnet Technology, p. 247, Oxford, UK, 1967.

[4] G. T. Danby, S. T. Lin, and J. W. Jackson, "Three dimensional properties of magnetic beam elements," in Proceedings of the 1967 Particle Accelerator Conference, Washington, D. C., March 1–3, 1967, ed. F. T. Howard (IEEE Trans. N. S. Vol. NS-14, No. 3, June 1967), p. 442.

[5] G. E. Lee-Whiting, "Measurement of quadrupole lens parameters," *Nucl. Instrum. Methods* 82 157 (1970).

[6] W. G. Davies, "The theory of the measurement of magnetic multipole fields with rotating coil magnetometers," *Nucl. Instrum. Methods* A311 399 (1992).

[7] S. Caspi, "Expressions of 3D magnetic harmonics and their applications to magnet ends," in the second special workshop on magnet simulations for particle accelerators, see reference [3.6].

[8] M. Venturini and A. Dragt, "Accurate computation of transfer maps from magnetic field data," *Nucl. Instrum. Methods (A)* 427 387 (1999).

[9] M. Venturini, *Lie Methods, Exact Map Computation, and the Problem of Dispersion in Space Charge Dominated Beams*, Ph. D. Dissertation, University of Maryland at College Park, Physics Department, 1998.

[10] C. Mitchell, *Calculation of Realistic Charged-Particle Transfer Maps*, Ph. D. Dissertation, University of Maryland at College Park, Physics Department, 2007.

[11] J. G. Wang, "Field distributions of injection chicane dipoles in SNS ring", in Proceedings of the 2005 Particle Accelerator Conference, Knoxville, TN, May 16–20, 2005, ed. C. Horak (IEEE Catalog Number 05CH37623C, 2005), p.3907.

[12] J. G. Wang, "Magnetic field distribution of injection chicane dipoles in Spallation Neutron Source accumulator ring," *Phys. Rev. ST Accel. Beams* 9, 012401 (2006). (http://prst-ab.aps.org/abstract/PRSTAB/v9/i1/e012401); also see J. G. Wang, SNS-NOTE-MAG-130, 2004.

[13] J. Simkin of Vector Fields Software of Cobham Technical Services, England, private communication.

[14] MATHEMATICA is a registered trademark of Wolfram Research, Inc., see, for example, Stephen Wolfram, "*The Mathematica Book*," 4th ed. (Wolfram Media/Cambridge University Press, 1999).

Chapter 5

PARTICLE OPTICS IN A SINGLE QUAD

Magnetic devices such as dipoles and quadrupoles in accelerator beam lines are usually treated as special lenses which can be characterized by particle optics or transfer maps. The optics of charged particles in magnetic quadrupoles has been well studied and the general information on the subject can be found in monographs [1–6]. In addition, many articles are specifically devoted to the analysis of magnetic fringe fields in a single quad [7–16]. Although the theory on magnetic fringe fields in a quadrupole has been well developed, it appears that their effects on the particle optics are often overlooked in practice. The difficulties are how to correctly derive the particle optics of quadrupoles from magnetic data through either simulation or measurement. Magnetic quadrupoles with small aspect ratios are often simply represented by "conventional hard edge models" obtained by calculating its magnetic length from the integrated gradient divided by the gradient around the magnet center. This in general is inaccurate for quads with significant fringe fields, as shown for an "all fringe" quad in a more recent paper [17].

The SNS accumulator ring contains magnetic quadrupoles all with large aperture and short steel length. Typical examples are the quadrupoles designated as 30Q58 and 30Q44 in a quad doublet assembly [18, 19], where 30 is the magnet aperture in centimeter and 58 and 44 are the iron yoke length in centimeter, respectively. The aspect ratios of these magnets are only 1.93 and 1.47. Significant fringe fields are expected. In this chapter, we investigate the particle optics of these two magnets and demonstrate the effect of magnetic fringe fields on particle dynamics [20]. We first build three-dimensional simulation models of these quads and study magnetic field distributions. The transfer matrices can be obtained directly from particle trajectory data. A more rigorous approach is to make the three-dimensional multipole expansion of their fields by following the techniques described in Chapter 4. The magnetic fringe fields are analyzed by the form factor theory, which leads to linear transfer matrices of the quads. These are compared with the solutions obtained directly from the trajectory equations. The hard edge models, as well as the lens parameters, are correctly derived from the matrix elements, and are compared with the conventional hard edge models. The third-order effect, which comes from the fringe fields such as the pseudo-octupole, is demonstrated by the rays on two symmetry planes.

5.1. Simulation Model of 30Q58 and Field Distributions

The quadrupole doublet assemblies are employed in the straight sections of the SNS ring for beam injection and extraction. In order to reduce the injection and extraction angles, the quadrupole yoke width in the x-direction is minimized. This creates the so-called "narrow quads." The quadrupoles have an aperture radius of R = 15.1232 cm and are driven by four conducting coils of 41 turns each. The four pole-tips are carefully optimized for good field quality, and their two ends are chamfered at a certain angle to minimize the dodecapole terms. In the quad doublet assembly, 30Q58 is energized as a focusing quad in the x-direction. The current density in its coil is set to 267.5694 A/cm^2, which corresponds to a magnet current of 810 A. This should be close to the 1 GeV operation point in the SNS ring. According to Eq. (2.29), the gradient of 30Q58 should be 364.94 G/cm.

The 30Q58 model in OPERA-3D/TOSCA simulation is shown in Figure 5.1. The center of the coordinate system coincides with the magnet geometric center and the cgs units are used. In the simulation the symmetry with respect to the three orthogonal planes across the coordinate axes is applied. Thus, only one eighth of the magnet needs to be processed in mesh generation and TOSCA solution. In this way, simulation time and model volumes are much reduced. In the OPERA-3D post processor a full magnet model is constructed and the symmetry with respect to z = 0 is always applied to simplify calculations.

The magnetic field distributions and qualities can be obtained from the OPERA-3D post-processor. Figure 5.2 shows the magnetic field component B_y on the positive x-axis at z = 0. The field appears very linear. The integral of this field from x = 0 to x = 10 cm is 18168.13 G-cm, yielding a gradient of 363.36 G/cm. This is only 0.43% smaller than that calculated from Eq. (2.29).

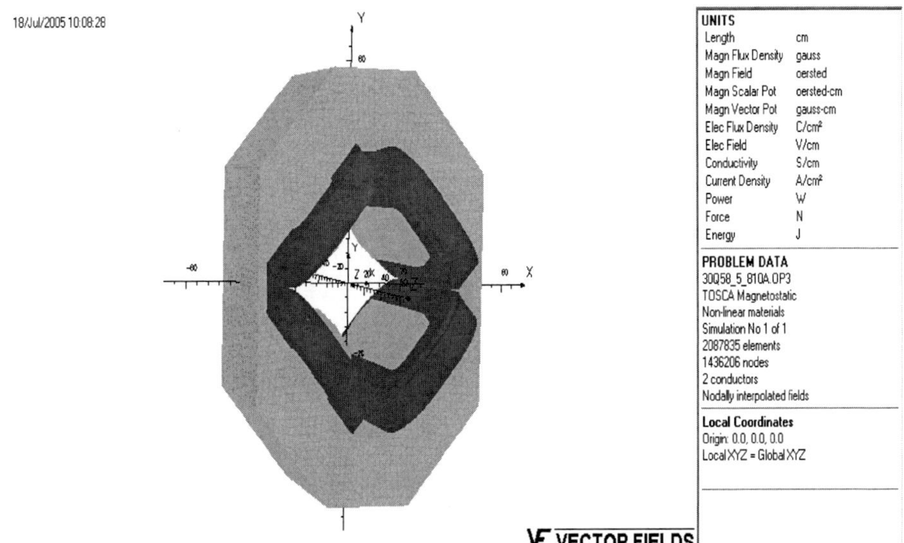

Figure 5.1. 30Q58 simulation model.

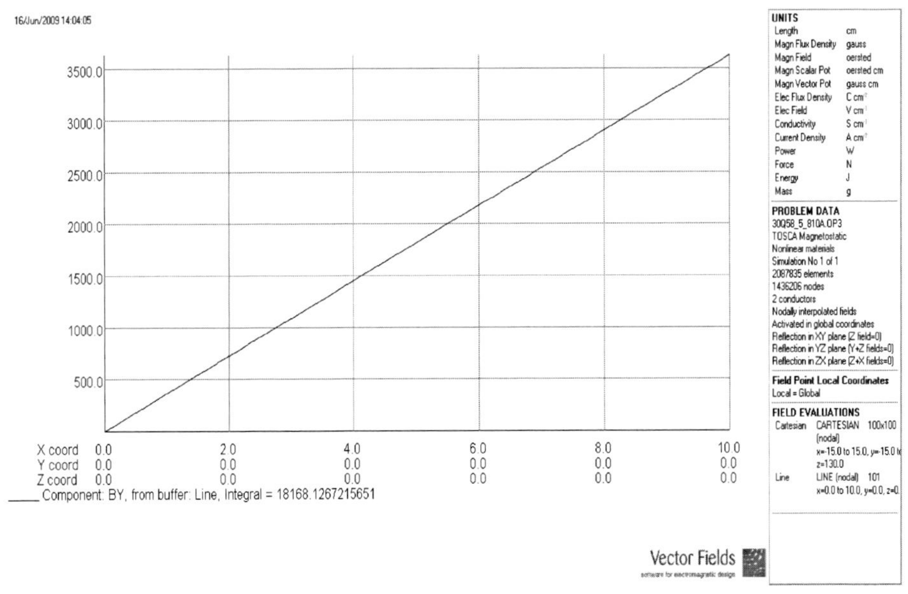

Figure 5.2. B_y vs. x at y = z = 0.

Figure 5.3 shows the magnetic field B_y versus z at x = 10 and y = 0. The fringe field extends far away. At z = 100 cm , the field B_y is still more than 0.5 G, and it drops to below 0.05 G at z = 130 cm. The field integral from z = 0 to z = 130 cm is 122852.95 G cm, resulting in an integrated gradient of about 12285.30 G for the half magnet. The field at z = 0 is 3629.96 G, leading to an approximate hard edge half-length of 33.84 cm. By employing the patch rotation method as described in Section 2.6, we obtain the integrated gradient of 2.4576 T for the full magnet. This number is very close to the result obtained from Figure 5.3. The integrated dodecapole, 20th pole, and 28th pole account for 2.43, 7.62, and 0.94 units, respectively. The other un-allowed harmonics such as sextupole, octupole, and decapole are negligibly small as expected.

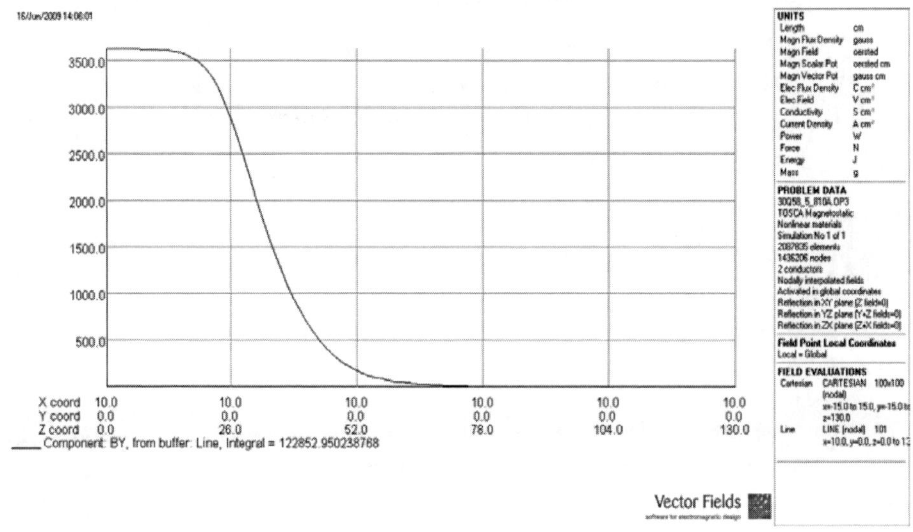

Figure 5.3. B_y vs. z at x = 10 and y = 0.

5.2 PARTICLE TRAJECTORIES AND TRANSFER MATRICES

OPERA-3D post processor has a built-in TRACK command for calculating trajectories of charged particles by directly integrating the equations of motion in simulated magnetic fields [21]. This provides a convenient way to calculate the transfer matrices of magnetic quadrupoles [22, 23]. For the x-z plane, the linear transfer matrix M_x of a magnetic quadrupole satisfies the relationship

$$\begin{pmatrix} x_2 \\ x_2' \end{pmatrix} = \mathbf{M}_x \begin{pmatrix} x_1 \\ x_1' \end{pmatrix} = \begin{bmatrix} M_{11} & M_{12} \\ M_{21} & M_{22} \end{bmatrix} \begin{pmatrix} x_1 \\ x_1' \end{pmatrix}. \tag{5.1}$$

where (x_1, x_1') and (x_2, x_2') are the phase space parameters at the magnet entrance and exit. We launch test particles with initial parameters (x_1, x_1') at the quad entrance, calculate their trajectories by the TRACK command, and obtain their final parameters (x_2, x_2') at the quad exit. The transfer matrix elements can be obtained from Eq. (5.1). Only two test particles are needed to fulfill the task. The simplest way is to pick one test particle with (x_1, $x_1' = 0$) and the other with ($x_1 = 0$, x_1'). From the first test particle, we obtain

$$\begin{cases} M_{11} = x_2 / x_1 \\ M_{21} = x_2' / x_1 \end{cases}, \tag{5.2a}$$

and from the second test particle, we have

$$\begin{cases} M_{12} = x_2 / x_1' \\ M_{22} = x_2' / x_1' \end{cases}. \tag{5.2b}$$

There is a similar relationship for the transfer matrix $\mathbf{M}_y$ for the y-z plane.

In the calculation for 30Q58, we define $z_1 = -130$ and $z_2 = 130$ cm as the entrance and exit points of the quad. This will include the entire fringe field. The first test particle starts at $x_1 = 2$, $y_1 = 0$, $z_1 = -130$ cm and $x_1' = y_1' = 0$, while the second test particle starts at $x_1 = y_1 = y_1' = 0$, $z_1 = -130$ cm and $x_1' = 0.010472$ ($\text{Arctan}(x_1') = 0.6^\circ$), which is set by the Euler's angles: phi = 0, theta = 0.6, and psi = 0. The particle energy is set to 1 GeV. The two trajectories calculated by the TRACK command are plotted in Figure 5.4. We then define a rectangular patch at $z_2 = 130$ cm. The patch is perpendicular to the quad axis and is large enough to intersect the two particles at z_2. The intersection of the calculated particle trajectories on the patch can be obtained in the TRACK command. This yields the particle positions and velocities, as well as (x_2, x_2'), at the quad exit. The transfer matrix $\mathbf{M}_x$ is then obtained as

$$\mathbf{M}_x = \begin{bmatrix} 0.4599 & 1.9120 \\ -0.4120 & 0.4585 \end{bmatrix}. \tag{5.3a}$$

The determinant of this matrix is 0.9985. Similarly, we obtain the transfer matrix $\mathbf{M}_y$ as

$$\mathbf{M}_y = \begin{bmatrix} 1.5937 & 3.3477 \\ 0.4604 & 1.5958 \end{bmatrix}. \tag{5.3b}$$

The determinant of this matrix is 1.0020. In the calculation of $\mathbf{M}_y$, the Euler's angles for the second particle should be phi = 90, theta = 0.6, and psi = 0. In an ideal quadrupole, the transfer matrix elements M_{11} and M_{22} should be the same, and the determinant of its transfer matrix should equal 1. These are only approximately true in the results (5.3a and 5.3b). The approximation comes from two aspects: field calculation accuracy in computation of the trajectories and nonlinear fields in the quad.

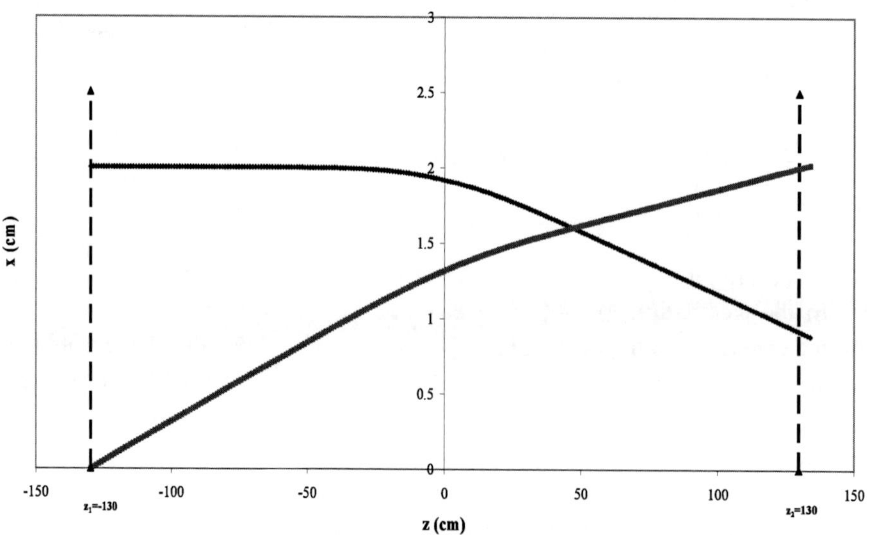

Figure 5.4. Particle trajectories on the x-z plane for calculating $\mathbf{M}_x$.

It should be noted that Eq. (5.1) represents a linear transfer matrix. However, the data from the particle trajectories contain both linear and nonlinear effects. In order to achieve more accurate results, the calculated particle trajectories should be as close to the quad axis as possible. Since the magnetic field in the vicinity of the quad axis is quite weak, it contains large relative errors. A particle trajectory too close to the axis can yield inaccurate data. Thus, a trade-off has to be made in simplified calculations. Figures 5.5(a) and 5.5(b) shows the results for two different field calculation methods at various initial particle positions. The corresponding numerical data are listed in Table 5.1. With the magnetic field calculated by the nodal interpolation method in OPERA-3D, the errors are large if the initial parameters are too small and the particle trajectories are too close to the quad axis. Thus, the determinants of the matrices deviate more from unity. On the other hand, the field calculation by the integration method in the code yields much more accuracy for particle trajectories. The smaller the initial parameters are, or the closer to the quad axis the particle trajectories are, the more accurate the transfer matrices are obtained. In the figures, the matrix elements M_{11} and M_{21} decreases monotonically when the initial particle position x_1 increases. This is caused by the nonlinearities of the quadrupole. From Table 5.1, the calculated data in the last row show that $M_{11} = M_{22}$ numerically up to the seventh decimal point and the determinant of the matrix is very close to 1.

Table 5.1 M_x elements from different calculation methods and initial parameters

x_1(cm) ($x_1' = 0$)	m_{11}	m_{21}(1/m)	theta (°) ($x_1 = 0$)	m_{12} (m)	m_{22}	Determinant
			Field calculation by nodal interpolation			
10	0.46070	-0.41158	3	1.91542	0.46039	1.00046
8	0.46064	-0.41154	2.4	1.91535	0.46051	1.00037
5	0.46073	-0.41136	1.5	1.91506	0.46049	0.999944
3	0.46049	-0.41151	0.9	1.91392	0.45981	0.99933
2	0.45992	-0.41196	0.6	1.91195	0.45850	0.9985
1	0.45671	-0.41449	0.3	1.90450	0.45297	0.9963
0.5	0.44866	-0.42061	0.15	1.89015	0.44191	0.9933
			Field calculation by integration			
10	0.460842	-0.411495	3	1.915697	0.460533	1.00053
5	0.460919	-0.411185	1.5	1.915864	0.460946	1.00023
2	0.461069	-0.411003	0.6	1.915929	0.461074	1.000039
1	0.4610915	-0.4109755	0.3	1.9159382	0.4610930	1.0000098
0.5	0.4610973	-0.4109687	0.15	1.9159406	0.4610976	1.0000025
0.25	0.46109870	-0.41096697	0.075	1.91594119	0.46109877	1.00000059
0.125	0.46109907	-0.41096653	0.0375	1.91594150	0.46109918	1.00000022
0.05	0.46109942	-0.41096623	0.0075	1.91594167	0.46109941	0.9999999984

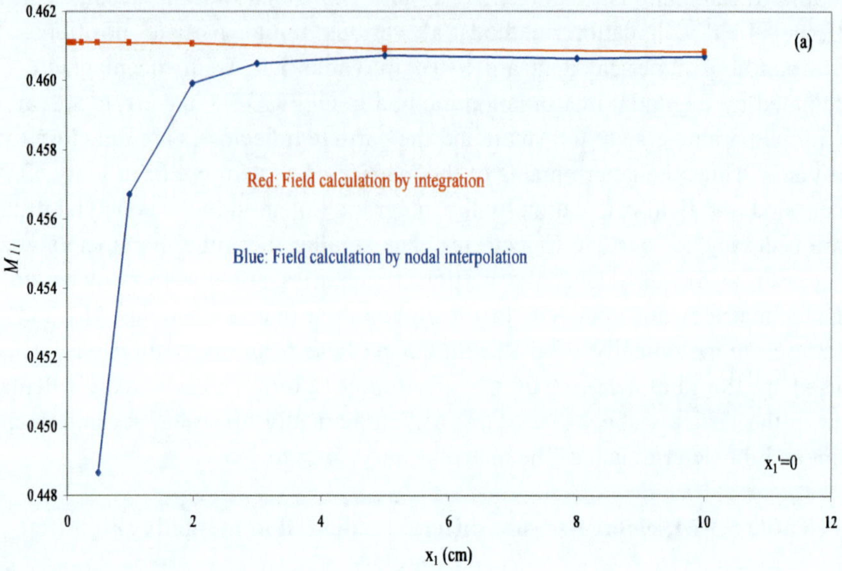

Figure 5.5. (a) M_{11} from various initial particle positions with two different field calculation methods.

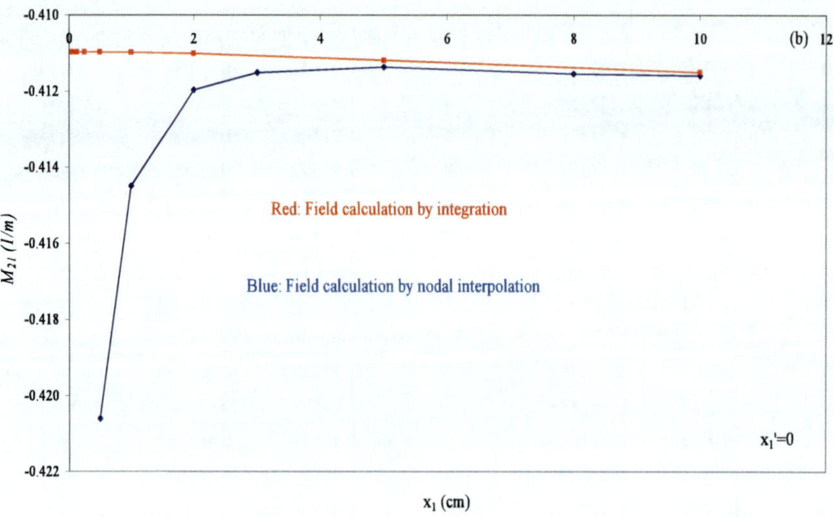

Figure 5.5. (b) M_{21} from various initial particle positions with two different field calculation methods.

Although the transfer matrix elements can be obtained with high accuracy from particle trajectory information if the field calculation method is set to the integration method, it always takes much longer computation time. It is also not known a priori what initial parameters can lead to good results. A more rigorous and systematic approach to linear transfer matrices is through three-dimensional multipole expansion and the integration of the equation of motion as described below.

5.3. THREE-DIMENSIONAL MULIPOLE EXPANSION AND LINEAR FOCUSING FUNCTION

The z-dependent field distribution, especially the linear focusing function of the quad, can be obtained by the techniques of three-dimensional multipole expansion of field data in simulation as described in Chapter 4. The expansion starts with the simulated radial field $B_r(R, \theta, z)$ on a cylindrical surface of radius R = 10 cm, which is co-axial with the magnet z-axis. The surface field B_r is calculated by the method of nodal interpolation in the expansion. A Fourier decomposition of $B_r(R, \theta, z)$ yields the harmonic contents of the field. For a reasonably well designed quadrupole in simulation, only the allowed terms (m = 2, 6, 10, 14, etc.) exist and their skew components can be neglected. In Figure 5.6, we plot only the normal quadrupole term (m = 2) of $B_r(R = 10, \theta, z)$. Further analyses lead to the generalized gradient $C_{2,s}(z)$ and its pseudo-multipoles.

The quadrupole field (m = 2 only) in a quadrupole magnet can be expressed up to third order by [24]

$$B_r(r,z) = 2C_{2,s}(z)r - \frac{1}{3}C_{2,s}''(z)r^3 + O(5), \tag{5.4a}$$

$$B_\theta(r,z) = 2C_{2,s}(z)r - \frac{1}{6}C_{2,s}''(z)r^3 + O(5), \tag{5.4b}$$

$$B_z(r,z) = 2C_{2,s}'(z)r^2 + O(4). \tag{5.4c}$$

The field component B_r(r, z) at r = R = 10 cm is also shown in Figure 5.6, where the blue curve is the linear term only, and the green curve includes the pseudo-octupole. The third order presentation of B_r(R, z) is a good approximation of the real field $B_2(R = 10, z)$ (the red curve in Figure 5.6). The effect of the pseudo-octupole on the particle optics will be described in Section 5.7.

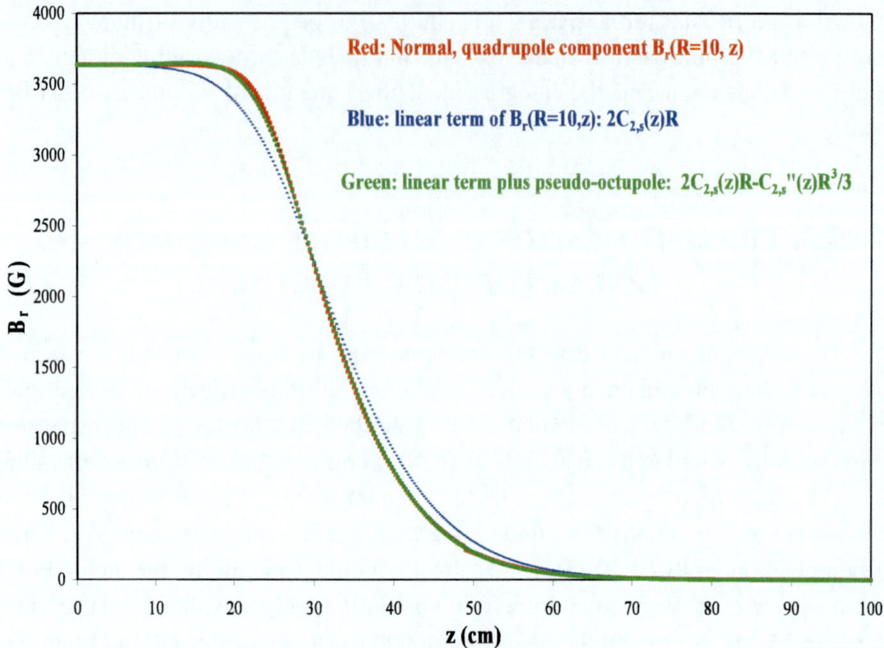

Figure 5.6. Normal quad term (m = 2) of radial component $B_r(R = 10, z)$ of 30Q58 and its constituents.

The linear focusing function k(z) of a quad is related to the normal, generalized linear gradient $C_{2,s}(z)$, which is an half of the quad gradient g(z) introduced in Section 2.5, by

$$k(z) = \frac{2C_{2,s}(z)}{B_\rho}, \tag{5.5}$$

where B_ρ is the magnetic rigidity of particles, which is 5.6574 T m for the proton at 1 GeV in the SNS ring. The linear focusing function k(z) contains all the information about magnetic fringe fields in the quad up to the first order. The right half of the linear focusing function of 30Q58 from this analysis is plotted in Figure 5.7. The reflection with respect to z = 0 is implied. It can be seen that the fringe field of the quad extends to a long distance. We cut off the plot at z = 1.3

m, beyond which the very small residual field is negligible. The average k(z) in the central region from z = 0 to z = 0.1 m is 0.64231 m^{-2}, corresponding to a quad gradient g_0 of 3.6338 T/m. The integration of k(z) from z = 0 to 1.3 m yields 0.21738 m^{-1}, which corresponds to an integrated gradient of 1.2298 T. The integrated gradient for the full quad from z = -1.3 to z = 1.3 m is 2.4596 T. This number, obtained from three-dimensional multipole expansion, can be compared with 2.4576 T, which is computed from the patch rotation method in Section 5.1. The difference is only 0.08%.

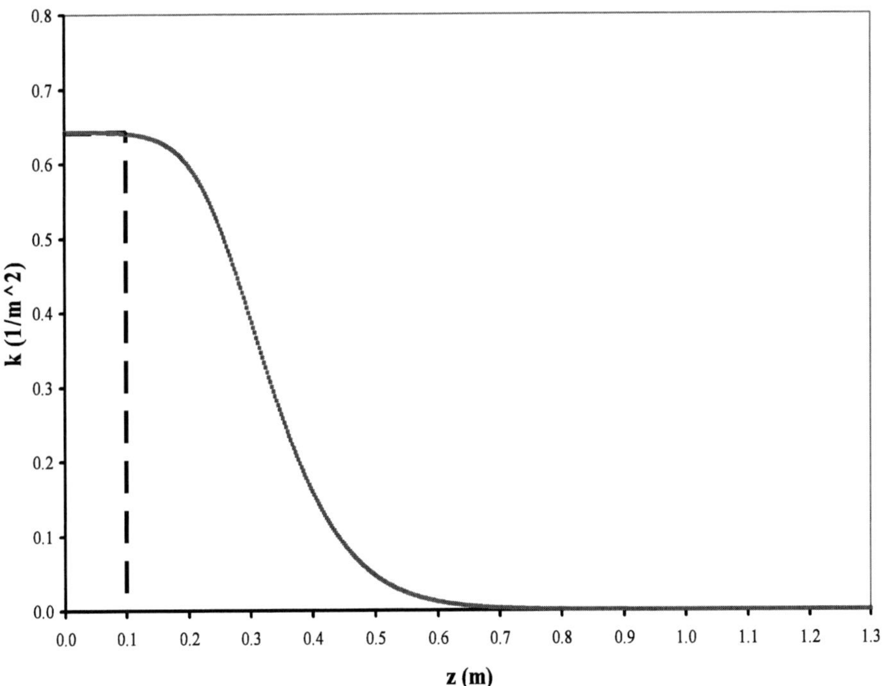

Figure 5.7. 30Q58 linear focusing function k(z).

5.4. FORM FACTOR THEORY ON MAGNETIC FRINGE FIELDS

The magnetic fringe field in a first-order theory of quadrupole lenses was studied long ago by G. E. Lee-Whiting [9]. In his theory, the focusing function is divided into two regions as indicated by the dashed line at z = 0.1 m in Figure 5.7: the central flat region is treated as an ideal hard edge quad with a constant focusing strength k_0; the fringe region is analyzed separately by normalizing its focusing strength to a dimensionless function as shown in Figure 5.8, where ζ is the normalized distance of the fringe region, $h(\zeta)$ is the normalized focusing strength, and $\xi = \kappa(\Delta z)$ with $\kappa = k_0^{1/2}$ and Δz as the total fringe length.

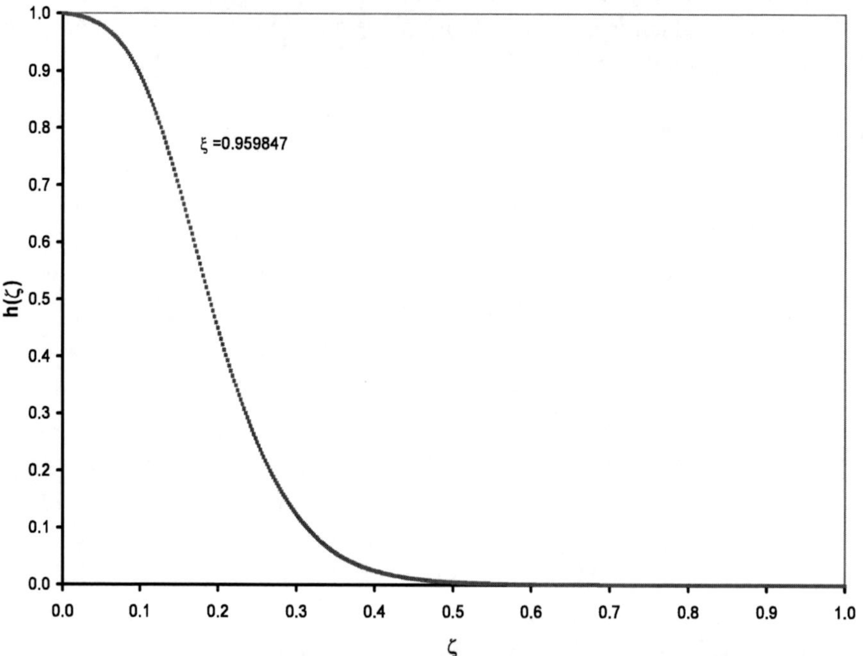

Figure 5.8. Normalized fringe of 30Q58.

By using the method of successive approximation, Lee-Whiting showed that the transfer matrices for the normalized fringe can be expressed as

$$\mathbf{m} = \begin{bmatrix} 1 \mp \varphi_2 \xi^2 + \phi_2 \xi^4 + \dots & \kappa^{-1}\xi\left(1 \mp \varphi_{12}\xi^2 + \phi_{12}\xi^4 + \dots\right) \\ \mp \kappa\xi\left(\varphi \mp \phi\xi^2 + \vartheta\xi^4 + \dots\right) & 1 \mp \varphi_1 \xi^2 + \phi_1 \xi^4 + \dots \end{bmatrix}. \tag{5.6}$$

Here the upper sign refers to the x-motion or focusing, and the lower sign to the y-motion or defocusing. The quantities φ_v are obtained by

$$\varphi_\nu = \int_0^1 h(\zeta)\Psi_1(\zeta)d\zeta, \tag{5.6a}$$

with $\Psi_1(\zeta) = 1, \zeta, (1-\zeta), \zeta(1-\zeta)$ for φ, φ_1, φ_2, φ_{12}, respectively. Similarly,

$$\phi_\nu = \int_0^1 h(\zeta)d\zeta \int_0^\zeta h(\zeta')(\zeta-\zeta')\Psi_2(\zeta,\zeta')d\zeta', \tag{5.6b}$$

with $\Psi_2(\zeta,\zeta') = 1, \zeta', (1-\zeta), \zeta'(1-\zeta)$ for ϕ, ϕ_1, ϕ_2, ϕ_{12}, respectively. And finally,

$$\vartheta = \int_0^1 h(\zeta)d\zeta \int_0^\zeta h(\zeta')(\zeta-\zeta')d\zeta' \int_0^{\zeta'} h(\zeta'')(\zeta'-\zeta'')d\zeta''. \tag{5.6c}$$

We compute **m** for the normalized fringe field in Figure 5.8 to obtain the matrix for the x-motion:

$$\mathbf{m} = \begin{bmatrix} 0.83732 & 1.17818 \\ -0.15186 & 0.97782 \end{bmatrix}.$$

This is the transfer matrix of the fringe field from z = 0.1 m to z = 1.3 m. The transfer matrix from z = -1.3 m to z = -0.1 m can be obtained by swapping the diagonal elements because of the rule of reflection with respect to the quad center. The transfer matrix from z = -0.1 m to z = 0.1 m is the one calculated from an ideal hard edge model. Concatenations of these three transfer matrices together yield the transfer matrix for the x-motion from $z_1 = -1.3$ m to $z_2 = 1.3$ m as

$$\mathbf{M}_x = \begin{bmatrix} 0.4591 & 1.9100 \\ -0.4108 & 0.4591 \end{bmatrix}. \tag{5.7a}$$

The determinant of this matrix is 0.9953. A similar computation for the y-motion leads to a transfer matrix from $z_1 = -1.3$ m to $z_2 = 1.3$ m as

$$\mathbf{M}_y = \begin{bmatrix} 1.5951 & 3.3495 \\ 0.4596 & 1.5951 \end{bmatrix}, \tag{5.7b}$$

and its determinant is 1.0049.

With the transfer matrices thus calculated, the form factor theory will lead to lens parameters and hard edge models of the quad as described in Section 5.6.

5.5. Linear Transfer Matrices from the Trajectory Equations

The form factor theory is very elegant for providing the physical insight of magnetic fringe fields in a quadrupole magnet. It also provides the formulation of the linear transfer matrix calculations. However, numerical computations of the multiple integrations according to Eqs. (5.6a, b, c) for a specific problem are not trivial. And the accuracy of the numerical results often remains in questions, which will be explained later. With modern computing technology such as the MATHAMATICA code [25], it should be much easier and more accurate to solve the same problem by direct numerical solutions of the particle trajectory equations. The linear trajectory equations of particle motion in a magnetic quadrupole are introduced in Section 2.5 and rewritten below

$$x''+k(z)x=0\,, \tag{5.8a}$$

$$y''-k(z)y=0\,. \tag{5.8b}$$

Here the first equation governs the focusing motion in the x-direction, while the second equation is for defocusing in the y-direction. The second-order differential equations can be easily solved numerically by two first-order differential equations of the following kind:

$$u_1'=u_2\ , \tag{5.9a}$$

$$u_2'=\pm k(z)u_1\ , \tag{5.9b}$$

where u represents either x or y, and (u_1 , u_2) forms a pair of phase space parameters. In numerical solutions we employ two sets of initial conditions. A ray with u_1 = 1 and u_2 = 0 at the initial position z_1 = -1.3 m yields the matrix components of $M_{11}(z)$ and $M_{21}(z)$, and a ray with u_1 = 0 and u_2 = 1 leads to the matrix elements of $M_{12}(z)$ and $M_{22}(z)$. The numerical integrations are performed through the quad exit position at z_2 = 1.3 m. The four matrix elements for the x-motion from z_1 = -1.3 m to z_2 = 1.3 m are plotted in Figure 5.9. These curves are also the principle ray traces. These calculations are much more straightforward and easier than those from the form factor theory.

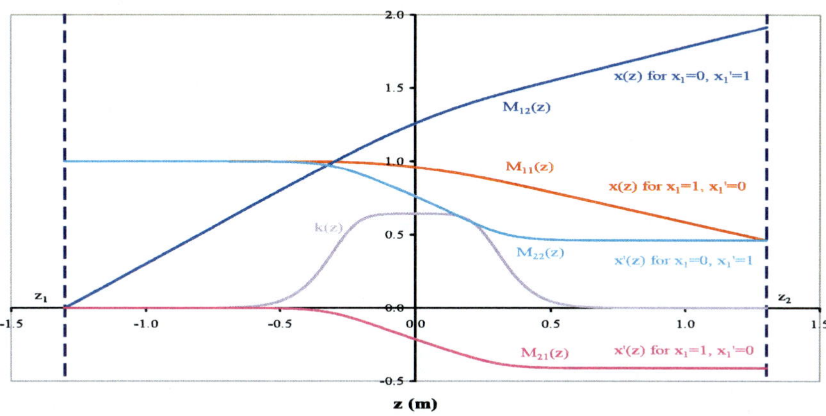

Figure 5.9. Matrix elements and ray traces for the x-motion in 30Q58.

The linear transfer matrix $\mathbf{M}_x$ of a focusing 30Q58 from $z_1 = -1.3$ m to $z_2 = 1.3$ m is evaluated from these traces at $z_2 = 1.3$ m as

$$\mathbf{M}_x = \begin{bmatrix} 0.46106 & 1.91587 \\ -0.41100 & 0.46106 \end{bmatrix}, \tag{5.10a}$$

and its determinant is

$$|\mathbf{M}_x| = 1.000000065.$$

For the y-motion in 30Q58, the four linear transfer matrix elements or ray traces are plotted in Figure 5.10. The linear transfer matrix from $z_1 = -1.3$ m to $z_2 = 1.3$ m is evaluated at $z_2 = 1.3$ m as

$$\mathbf{M}_y = \begin{bmatrix} 1.59235 & 3.34248 \\ 0.45941 & 1.59235 \end{bmatrix}, \tag{5.10b}$$

with a determinant

$$|\mathbf{M}_y| = 1.00000043.$$

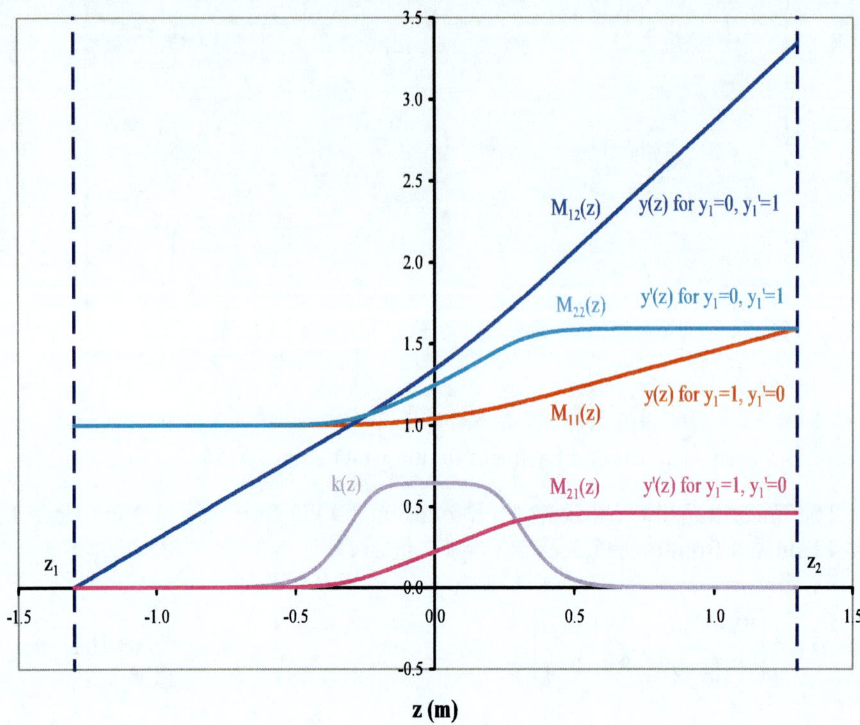

Figure 5.10. Matrix elements and ray traces for the y-motion in 30Q58.

The matrix elements of **M** in (5.10a, b) are very close to that in (5.7a, b) and (5.3a, b). Thus, we see the good agreement among the three different methods: the numerical solutions of the first-order trajectory equations, the form factor theory, and the test particle trajectory data. However, the determinants of these matrices in (5.10a, b) are much closer to unity than those in (5.7a, b), indicating that the solutions from the trajectory equations are more accurate than those from the form factor theory. This is due to the fact that any transfer matrix of a Hamiltonian system is symplectic, and the determinant of a symplectic matrix should be equal to 1. The larger errors with the form factor theory not only come from more complicated numerical computations, but also from the fact that the magnetic fringe field in 30Q58 is relatively too large in comparison with that in the central flat region, thus more terms of the iterated-integrations are required in Eqs. (5.6a, b, c).

5.6. LENS PARAMETERS AND HARD EDGE MODELS

With the linear transfer matrices in (5.10) or (5.7) or (5.3), the four lens parameters of the quad can be obtained. The focusing length is simply expressed as

$$f = f_1 = f_2 = -\frac{1}{M_{21}}. \tag{5.11}$$

And the principle plane distance is

$$d_1 = -\frac{1-M_{22}}{M_{21}} - l_1, \tag{5.12a}$$

$$d_2 = -\frac{1-M_{11}}{M_{21}} - l_2. \tag{5.12b}$$

Here l_1 and l_2 are the distances from the lens center to the entrance and exit, respectively. For a symmetric lens such as a single 30Q58, $l_1 = l_2 = l$, and $M_{22} = M_{11}$, thus, $d_1 = d_2 = d$. We use more accurate matrices (5.10) for numerical calculations of the lens parameters, as listed in Table 5.2.

A symmetric lens has only two independent parameters (f, d). A symmetric hard edge model also has two independent parameters (L, k_0). Therefore, it is possible to have an equivalent hard edge model for each symmetric lens. The equivalence is valid only in the sense that the hard edge model should yield the same transfer matrix from the lens entrance to the exit. Thus, we require the following relationship

$$\mathbf{M} = \mathbf{DHD}. \tag{5.13}$$

Here **M** is the transfer matrix of 30Q58 found above, and

$$\mathbf{D} = \begin{bmatrix} 1 & l - L/2 \\ 0 & 1 \end{bmatrix}$$

is the transfer matrix for a drift distance of (l - L/2). The transfer matrix for a hard edge model is

$$\mathbf{H} = \begin{bmatrix} C & S/\kappa \\ \mp\kappa S & C \end{bmatrix},$$

where C and S are cos(φ) and sin(φ) for a focusing quad, cosh(φ) and sinh(φ) for a defocusing quad, respectively, φ is the effective lens focusing strength, and kappa, κ, is the square root of the constant focusing function k_0. The upper sign (minus) in H_{21} is for focusing and the lower sign (plus) for defocusing. This relationship leads to a transcendental equation for φ [1, 9]:

$$M_{11} - M_{21}l = C \pm \frac{1}{2}\varphi S. \tag{5.14a}$$

The other hard edge parameters can be found as

$$\kappa = \frac{M_{21}}{\mp S}, \tag{5.14b}$$

$$L = \frac{\varphi}{\kappa}. \tag{5.14c}$$

Equation (5.14a) can be solved numerically to obtain the effective lens strength φ, and further to get κ, k_0, and L from Eqs (5.14b) and (5.14c). The hard edge parameters thus calculated for 30Q58 are also listed in Table 5.2.

The various hard edge models for 30Q58 are plotted in Figure 5.11. The linear focusing function k(z) is also included. In the "conventional hard edge model", its height equals the central strength of the quad, and its effective length is calculated from the integrated gradient divided by the gradient around the quad center. The first-order hard edge models come from rigorous computations based on the first-order trajectory equations and the resulting linear transfer matrices.

The plot in Figure 5.11 and the numerical data in Table 5.2 show clearly the differences between the conventional hard edge model and the first-order hard edge models. First, the hard edge parameters k_0, L, and k_0*L in the first-order focusing and defocusing models are slightly different. Thus, we need in principle two different hard edge models for focusing and defocusing, rather than one as in the conventional treatment. This may be very inconvenient in applying these models in the lattice design codes such as ORBIT [26], TRANSPORT [27], and MAD [28]. For many applications, we may still use one first-order hard edge model with the mean values of k_0 and L between the focusing and defocusing models as an approximation. Second, the hard edge height is no longer equal to that at the quad center, as in the case of the conventional model. In 30Q58, the mean hard edge height k_0 is about 15% smaller than that in the conventional model. Accordingly, the mean hard edge length L is longer than that in the conventional model by roughly the same amount. Third, the product of the hard edge height and length is not quite the same as the integrated focusing function. The relationship is held only for the geometric mean value between the first-order focusing and defocusing models. Fourth, as a result of all the above differences, the lens parameters also differ from that in the conventional hard edge model. For instance, the effective lens strength φ in the first-order hard edge model of 30Q58 is 2.2^o more than that in the conventional hard edge model, which would result in different focusing lengths and tunes. The focusing length of 30Q58 predicted by the first-order hard edge models differs from that of the conventional model by about 0.7%. All these differences come from the magnetic fringe field effect of the quad.

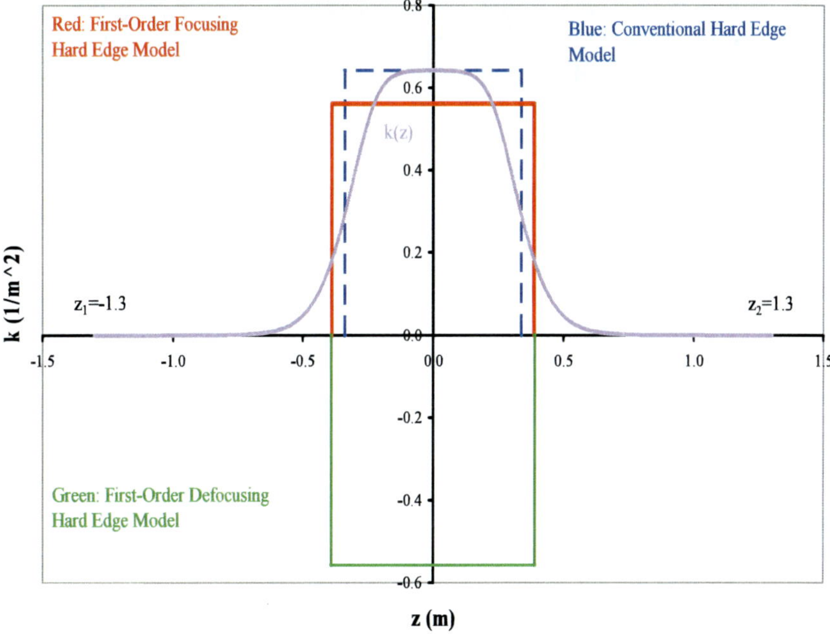

Figure 5.11. Hard edge models for 30Q58.

Table 5.2 30Q58 hard edge models and lens parameters

	Conventional	1st-order Focusing	1st-order Defocusing
L (m)	0.6769	0.7757	0.7794
k_0 (1/m^2)	0.6423	0.5609	0.5575
k_0*L (1/m)	0.4348	0.4351	0.4345
φ (degree)	31.08	33.28	33.34
d (cm)		1.129	1.064
f_x (m)	2.417	2.433	
f_y (m)	-2.191		-2.177

5.7. Third-Order Aberrations

The quadrupole (m = 2) field in a quadrupole magnet up to the third order in r is given in Eqs. (5.4a, b, c), where the second terms are so called pseudo-octupole. The pseudo-octupole has an azimuthal dependence of m = 2 and a radial dependence of r^3. The pseudo-multipoles are produced by magnetic fringe fields. If the generalized gradient $C_{2,s}$ has no z-dependence, its higher derivatives vanish, and the pseudo-multipoles disappear. This would be a two-dimensional field. Figure 5.6 shows the radial component $B_r(R, z)$ of 30Q58 at R = 10 cm and its first two constituents: linear gradient and pseudo-octupole. It is clear that the linear term alone is not accurate for $B_r(R, z)$. A third-order representation, which includes the linear gradient and the pseudo-octupole only, is usually a good approximation of the real field.

The third-order aberrations of a quadrupole magnet with fringes have been extensively studied [1–4, 10–15]. In this case, the particle optics can be expressed by

$$\mathbf{x_2} = \mathbf{M}\mathbf{x_1} + \Delta\mathbf{x}\,, \tag{5.15}$$

where $\mathbf{x}_1$ and $\mathbf{x}_2$ are the phase space vectors for the x-motion at the entrance z_1 and exit z_2, $\mathbf{M}$ is the first-order transfer matrix obtained before, and $\Delta\mathbf{x}$ is the third-order aberration vector found from

$$\Delta \mathbf{x} = \begin{bmatrix} P_{10} & P_{11} & P_{12} & P_{13} & Q_{10} & Q_{11} & Q_{12} & R_{10} & R_{11} & R_{12} \\ P_{20} & P_{21} & P_{22} & P_{23} & Q_{20} & Q_{21} & Q_{22} & R_{20} & R_{21} & R_{22} \end{bmatrix} \cdot$$

$$\bullet \begin{bmatrix} x_1^3 & x_1^2 x'_1 & x_1 x'^2_1 & x'^3_1 & x_1 y_1^2 & x_1 y_1 y'_1 & x_1 y'^2_1 & x'_1 y_1^2 & x'_1 y_1 y'_1 & x'_1 y'^2_1 \end{bmatrix}^T \tag{5.16}$$

The first term in the RHS of Eq. (5.16) is the aberration coefficient matrix, and the second term is a vector for the initial conditions. There is a similar expression for the y-motion. Thus, there are forty aberration coefficients, though not all independent due to the quadrupole symmetry. The third-order differential equations of motion can be obtained by employing the field expressions in Eq. (5.4) and keeping all the dynamic variable terms up to a third order:

$$x'' + kx = k(x'yy' - \frac{3}{2}xx'^2 - \frac{1}{2}xy'^2) + k'xyy' + \frac{1}{12}k''(3xy^2 + x^3), \tag{5.17a}$$

$$y'' - ky = -k(y'xx' - \frac{3}{2}yy'^2 - \frac{1}{2}yx'^2) - k'yxx' - \frac{1}{12}k''(3yx^2 + y^3). \tag{5.17b}$$

Equations (5.17a) and (5.17b) are usually solved by replacing x, x′, y, y′ on the right hand side by the appropriate first-order solutions. Here we only illustrate the third-order effects for particles on two symmetry planes and compare the results with the first-order theory. It can be shown that the third-order aberration vector taking into account x_1^3 only (P_{10}, P_{20} terms) can be found from

$$\Delta x'' + k\Delta x = (-\frac{3}{2}kM_{11}M_{21}^2 + \frac{1}{12}k''M_{11}^3)x_1^3; \tag{5.18a}$$

and the third-order aberration vector taking into account $x_1'^3$ only (P_{13}, P_{23} terms) can be found from

$$\Delta x'' + k\Delta x = (-\frac{3}{2}kM_{12}M_{22}^2 + \frac{1}{12}k''M_{12}^3)x'^3_1. \tag{5.18b}$$

In Figure 5.12 we plot the numerical solutions of Eqs. (5.18a, b), which are the third-order aberrations on the x-plane through 30Q58. The corresponding first-order rays result from two cases: one with the initial condition of $x_1 = 0.1$ m and

$x_1' = 0$ and the other with $x_1 = 0$ and $x_1' = 0.075$. The computed particle parameters and the aberration coefficients are listed in Table 5.3, where Δx(%) and Δx′(%) are the relative changes with respect to the linear optics parameters at the exit $z_2 = 1.3$ m. For the first ray, the nonlinear field changes x_2 and x_2' by -0.17% and 0.22%, respectively; and for the second ray the changes in x_2 and x_2' are -0.22% and -0.46%, respectively.

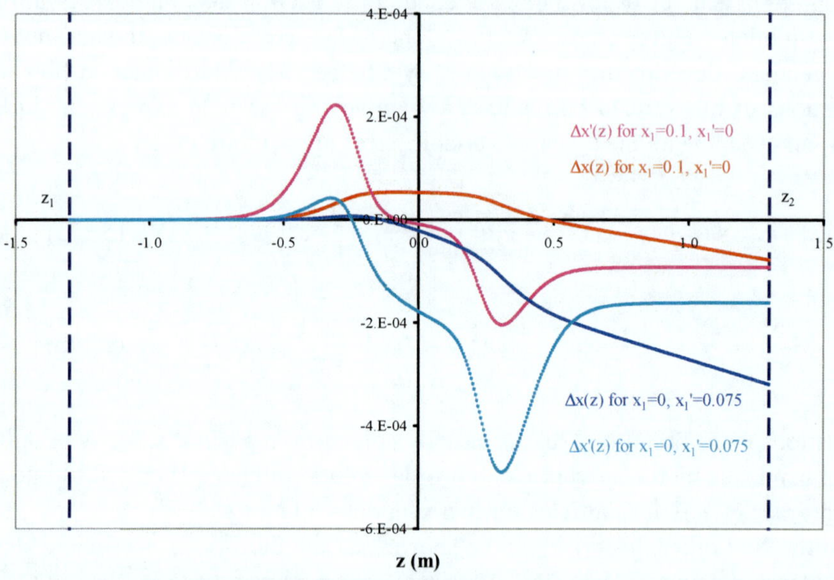

Figure 5.12. Third-order aberrations on the x-plane in 30Q58.

Table 5.3 Numerical data for third-order effect in 30Q58

Focusing Plane			Defocusing Plane		
Initial x_1(m)	0.1	0	Initial y_1(m)	0.1	0
Initial x_1'	0	0.075	Initial y_1'	0	0.075
1st-order x_2(m)	0.04611	0.14369	1st-order y_2(m)	0.15924	0.25069
1st-order x_2'	-0.04110	0.03458	1st-order y_2'	0.04594	0.11943
Δx (m)	-7.73E-05	-3.19E-04	Δy (m)	-1.80E-05	6.95E-04
Δx'	-9.17E-05	-1.60E-04	Δy'	-5.40E-05	3.87E-04
Δx (%)	-0.17%	-0.22%	Δy (%)	-0.011%	0.28%

Focusing Plane			Defocusing Plane		
$\Delta x'$ (%)	0.22%	-0.46%	$\Delta y'$ (%)	-0.12%	0.32%
P_{10} (m^{-2})	-0.0773		P_{10} (m^{-2})	-0.0180	
P_{20} (m^{-3})	-0.0917		P_{20} (m^{-3})	-0.0540	
P_{13} (m)		-0.756	P_{13} (m)		1.647
P_{23}		-0.379	P_{23}		0.917

For the y-motion in 30Q58, the third-order aberrations for the two special rays can be found from the following equations:

$$\Delta y''-k\Delta y=(\frac{3}{2}kM_{11}M_{21}^{2}-\frac{1}{12}k''M_{11}^{3})y_{1}^{3}, \tag{5.19a}$$

$$\Delta y''-k\Delta y=(\frac{3}{2}kM_{12}M_{22}^{2}-\frac{1}{12}k''M_{12}^{3})y_{1}'^{3}. \tag{5.19b}$$

Figure 5.13 plots the third-order aberrations on the y-plane in 30Q58 for two rays with different initial conditions. The numerical values are also listed in Table 5.3. For the first ray, the nonlinear field changes y_2 and y_2' by -0.011% and -0.12%, respectively; for the second ray the changes in y_2 and y_2' are 0.28% and 0.32%, respectively.

The analysis of third-order aberrations in a quadrupole magnet as presented above illustrates the principle of solving a nonlinear problem. It has shown the complexity of the problem and it would be difficult and time consuming with this approach to compute all the aberration coefficients. There are more advanced techniques, which can handle even higher-order aberrations. One of them is to employ the Lie algebra method [29, 30] to generate transfer maps to include high-order nonlinear terms. In this approach, three-dimensional multipole expansion is made for the magnetic vector potential based on magnet simulation data [31, 32]. A beam dynamics code MARYLIE [33] is employed to generate transfer maps through fifth order. A future code, MARYLIE 7.1, is currently under development, which will carry out all the computations through seventh order. A recent development of automating the computation of quadrupole transfer maps to the third order by using electromagnetic field solutions and MARYLIE code is reported in [34]. Another technique is to employ the differential algebra method [35–37], in which computations can be done through the code COSY or COSY INFINITY to arbitrary high order.

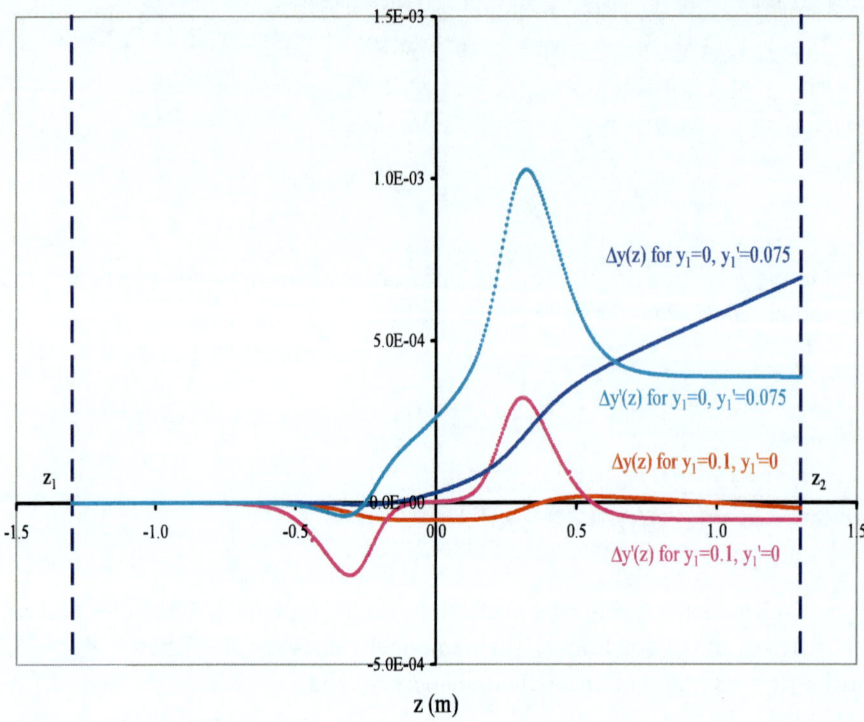

Figure 5.13. Third-order aberrations on the y-plane in 30Q58.

5.8. Particle Optics in 30Q44

Magnetic quadrupole 30Q44 in the SNS ring quad doublet assembly is used as a defocusing quad on the x-plane. It is energized at -860 A for 1 GeV operation. The linear focusing function k(z) of a single 30Q44 obtained from simulation and three-dimensional multipole expansion is plotted in Figure 5.14, where the reflection with respect to z = 0 is implied. The computation of particle optics in 30Q44 follows the same procedure as for 30Q58. We first calculate the linear transfer matrices according to Eq. (5.9). The lens parameters and the hard edge models are then computed. The third-order analysis yields the aberration coefficients in the two symmetry planes. We skip the detailed steps and plots, and only list the numerical results in Tables 5.4 and 5.5.

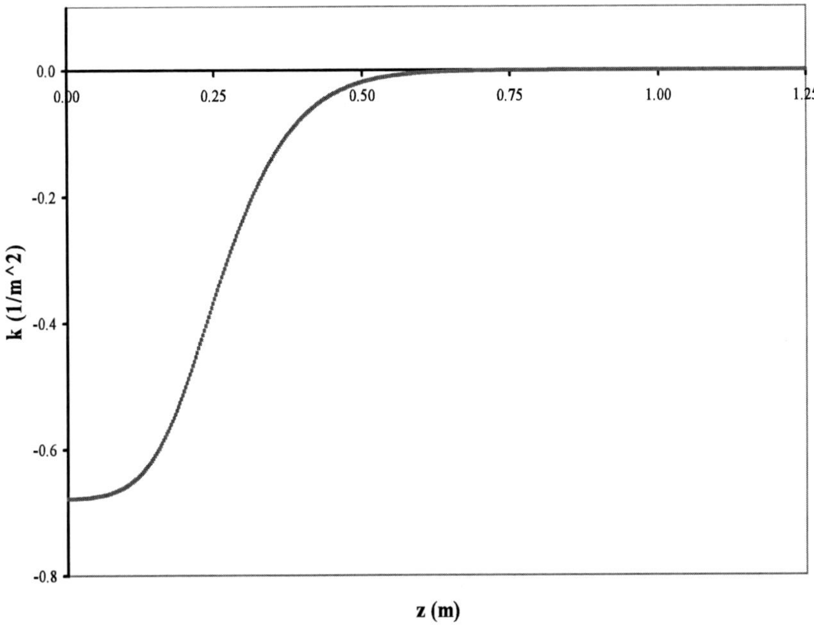

Figure 5.14. Linear focusing function of 30Q44.

Table 5.4. 30Q44 hard edge models, lens parameters, and matrix elements

	Conventional	1st-order Focusing	1st-order Defocusing
L (m)	0.5407	0.6605	0.6635
k_0 ($1/m^2$)	-0.6770	0.5547	-0.5513
k_0*L (1/m)	-0.3661	0.3664	-0.3658
φ (degree)	25.49	28.18	28.23
d (cm)		0.683	0.655
f_x (m)	2.824	2.843	
f_y (m)	-2.644		-2.626
Transfer matrix elements from z_1 = -1.25 m to z_2 = 1.25 m			
M_{11}		0.5579	1.4735
M_{12} (m)		1.9580	3.0757
M_{21} (1/m)		-0.3518	0.3808
M_{22}		0.5579	1.4735

Again, the numerical data in Table 5.4 show the differences between the conventional hard edge model and the first-order hard edge models for 30Q44. The mean hard edge height k_0 is about 22% smaller than that in the conventional model. Accordingly, the mean hard edge length L is longer than that in the conventional model by roughly the same amount. The effective lens strength φ in the first-order hard edge models is 2.7° more than that in the conventional hard edge model. The focusing lengths predicted by the first-order hard edge models differ from that of the conventional model by about 0.7%. The particle optics will further deviate from the prediction of conventional hard edge models by magnetic interference as described in the next chapter.

Table 5.5 30Q44 third-order aberration parameters

Defocusing Plane			Focusing Plane		
Initial x_1(m)	0.1	0	Initial y_1(m)	0.1	0
Initial x_1'	0	0.075	Initial y_1'	0	0.075
1st-order x_2(m)	0.0558	0.1469	1st-order y_2(m)	0.1474	0.2307
1st-order x_2'	-0.0352	0.0418	1st-order y_2'	0.0381	0.1105
Δx (m)	-6.30E-05	-2.73E-04	Δy (m)	-2.78E-05	4.91E-04
Δx'	-7.49E-05	-1.45E-04	Δy'	-5.09E-05	2.78E-04
Δx (%)	-0.11%	-0.19%	Δy (%)	-0.019%	0.21%
Δx' (%)	0.21%	-0.35%	Δy' (%)	-0.13%	0.25%
P_{10} (m^{-2})	-0.0630		P_{10} (m^{-2})	-0.0278	
P_{20} (m^{-3})	-0.0749		P_{20} (m^{-3})	-0.0509	
P_{13} (m)		-0.647	P_{13} (m)		1.164
P_{23}		-0.343	P_{23}		0.658

REFERENCES

[1] K. G. Steffen, *High Energy Beam Optics*, John Wiley and Sons, Inc., New York, 1965.

[2] P. Grivet, *Electron Optics*, Pergamon Press Ltd, London, 1965.

[3] P. W. Hawkes, *Quadrupole Optics*, Springer-Verlag, Heidelberg, 1966.

[4] H. Wollnik, *Optics of Charged Particles*, Academic Press, New York, 1987.

[5] D. C. Carey, *The Optics of Charged Particle Beams,* Harwood Academic Publishers, Paris, 1987.

[6] M. Reiser, *Theory and Design of Charged Particle Beams*, John Wiley and Sons, Inc., New York, 1994.

[7] H. Wollnik and H. Ewald, "The influence of magnetic and electric fringe fields on the trajectories of charged particles," *Nucl. Instrum. Methods* 36 93 (1965).

[8] H. Hubner and H. Wollnik, "The design of magnetic field clamps," *Nucl. Instrum. Methods* 86 141 (1970).

[9] G. E. Lee-Whiting, "End effects in first-order theory of quadrupole lenses," *Nucl. Instrum. Methods* 76 305 (1969).

[10] G. E. Lee-Whiting, "Third-order aberrations of a magnetic quadrupole lens," *Nucl. Instrum. Methods* 83 232 (1970).

[11] H. Matsuda and H. Wollnik, "Third order transfer matrices of the fringe field of an inhomogeneous magnet," *Nucl. Instrum. Methods* 77 283 (1970).

[12] H. Matsuda and H. Wollnik, "Third order transfer matrices for the fringe field of magnetic and electrostatic quadrupole lenses," *Nucl. Instrum. Methods* 103 117 (1972).

[13] J. Irwin and Chun-Xi Wang, "Explicit soft fringe maps of a quadrupole," in Proceedings of the 1995 Particle Accelerator Conference, Dallas, TX, May 1–5, 1995, ed. L. Gennari (IEEE Piscataway, NJ, 1996), p. 2376.

[14] R. Baartman, "Intrinsic third order aberrations in Electrostatic and magnetic quadrupoles," in Proceedings of the 1997 Particle Accelerator Conference, Vancouver, B. C., Canada, May 12–16, 1997 (IEEE Piscataway, NJ, 1998), p. 1415.

[15] M. Venturini, "Scaling of third-order quadrupole aberrations with fringe field extension," in Proceedings of the 1999 Particle Accelerator Conference, New York, March 29–April 2, 1999, eds. A. Luccio and W. Mackay (IEEE Piscataway, NJ, 1999), p.1590.

[16] R. Baartman and D. Kaltchev, "Short quadrupole parameterization," in Proceedings of the 2007 Particle Accelerator Conference, Albuquerque, NM, June 25–29, 2007 (IEEE Catalog Number: 07CH37866, 2007), p. 3229.

[17] S. Bernal, H. Li, R. A. Kishek, B. Quinn, M. Walter, M. Reiser, and P. G. O'Shea, C. K. Allen, "RMS envelope matching of electron beams from "zero" current to extreme space charge in a fixed lattice of short magnets," *Phys. Rev. ST Accel. Beams* 9, 064202 (2006). (http://prst-ab.aps.org/abstract/PRSTAB/v9/i6/e064202)

[18] N. Tsoupas, J. Brodowski, W. Meng, J. Wei, Y. Y. Lee, J. Tuozzolo, "A Large-aperture narrow quadrupole for the SNS accumulator ring," in Proceedings of the Eighth European Particle Accelerator Conference, Paris,

France, June 3–7, 2002, eds. T. Garvey et al. (EPS-IGA and CERN, 2002), p. 1106.

[19] N. Tsoupas, J. Jackson, Y.Y. Lee, D. Raparia, J. Wei, "Magnetic field calculations for a large aperture narrow quadrupole," in Proceedings of the 2003 Particle Accelerator Conference, Portland, OR, May 12–16, 2003, eds. J. Chew, P. Lucas, and S. Webber (IEEE Piscataway, NJ, 2003), p. 2153.

[20] J. G. Wang, "Particle optics of quadrupole doublet magnets in Spallation Neutron Source accumulator ring," *Phys. Rev. ST Accel. Beams* 9, 122401 (2006). (http://prst-ab.aps.org/abstract/PRSTAB/v9/i12/e122401); also see J. G. Wang, SNS-NOTE-MAG-168, July 12, 2006.

[21] OPERA-3D User Guide and Reference Manual, Vector Fields Software of Cobham Technical Services, Oxford, England; also see http://www.vectorfields.com/

[22] E. R. Close, "Generation of first and second order transformation elements from a given magnetic field," *Nucl. Instrum. Methods* 89 205 (1970).

[23] B. Langenbeck, "Getting ion optical transform matrices from TOSCA/OPERA calculations," in Proceedings of the 1992 Vector Fields users meeting, Louvain-la-Neuve, Belgium, 1992.

[24] J. G. Wang, "Magnetic field distribution of injection chicane dipoles in Spallation Neutron Source accumulator ring," *Phys. Rev. ST Accel. Beams* 9, 012401 (2006). (http://prst-ab.aps.org/abstract/PRSTAB/v9/i1/e012401); also see J. G. Wang, SNS-NOTE-MAG-130, 2004.

[25] MATHEMATICA is a registered trademark of Wolfram Research, Inc., see, for example, Stephen Wolfram, "*The Mathematica Book*," 4th ed. (Wolfram Media/Cambridge University Press, 1999).

[26] J. A. Holmes, S. Cousineau, V. V. Danilov, S. Henderson, A. Shishlo, Y. Sato, W. Chou, L. Michelotti, and F. Ostiguy, "ORBIT: Beam dynamics calculations for high intensity rings," *in The ICFA Beam Dynamics Newsletter*, 30 (2003).

[27] D. C. Carey, K. L. Brown, and F. Rothacker, *Third-Order TRANSPORT with MAD Input, A Computer Program for Designing Charged Particle Beam Transport Systems,* SLAC-R-530, Fermilab-pub-98-310, UC-414, 1998.

[28] H. Grote and F. C. Iselin, *The MAD Program (Methodical Accelerator Design),* CERN/SL/90-13 (AP), 1990.

[29] A. J. Dragt, *Lie Methods for Nonlinear Dynamics with Applications to Accelerator Physics* (University of Maryland, College Park, Physics Department Report, 2007.)

[30] A. J. Dragt, "A method of transfer maps for linear and nonlinear beam elements," *IEEE Transactions on Nuclear Science* NS-26 (3), 3601 (1979).

[31] M. Venturini, *Lie Methods, Exact Map Computation, and the Problem of Dispersion in Space Charge Dominated Beams*, Ph. D. Dissertation, University of Maryland at College Park, Physics Department, 1998.

[32] C. Mitchell, *Calculation of Realistic Charged-Particle Transfer Maps*, Ph. D. Dissertation, University of Maryland at College Park, Physics Department, 2007.

[33] MARYLIE code and manual, see http://www.physics.umd.edu/dsat/

[34] G. H. Gillespie, B. W. Hill, J. F. DeFord, and B. Held, "Automating the computation of quadrupole transfer maps and matrices utilizing electromagnetic field solutions," in Proceedings of the 2009 Particle Accelerator Conference, Vancouver, B. C., Canada, May 4–8, 2009, ed. M. Comyn, TH6PFP077.

[35] M. Berz and H. Wollnik, "Analytic solution of single-particle equations of motion to high orders: the formula manipulator HAMILTON and the fifth-order code COSY 5.0," in AIP Conference Proceedings 177: Linear Accelerator and Beam Optics Codes, La Jolla Institute 1988, ed. C. R. Eminhizer (New York 1988), p. 301.

[36] M. Berz, "Differential algebraic treatment of beam dynamics to very high orders," *Particle Accelerators* 24 (2), 109 (1989).

[37] G. H. Hoffstatter and M. Berz, "Symplectic scaling of transfer maps including fringe fields," *Phys. Rev. E* 54 (6), November 1996, p. 5664.

Chapter 6

MAGNETIC INTERFERENCE BETWEEN TWO MAGNETS

Small particle accelerator rings are usually densely packed due to very limited space among lattice components. Magnetic fields between adjacent magnets often overlap or are affected by surrounding materials. This leads to magnetic interference, which may play important role in machine commissioning and operation. The studies of magnetic interference in the past has been limited to simulations or measurements of the changes in the integrated main fields and their harmonics of a magnet due to its neighbors [1–4]. These changes are a measure of the strength of magnetic interference. However, they neither provide information about the effect on particle motion, nor can they be used to compensate the integrated strength of magnets in an existing lattice consisting of conventional hard edge models. In reference [5] we developed more rigorous analyses of particle optics under the influence of magnetic interference in a quadrupole assembly, and this material is presented in this Chapter.

With the example of the magnetic quadrupole 30Q58 presented in Chapter 5, a magnetic dipole corrector 41CD30 is placed closely around the quad. We analyze the magnetic interference between these two magnets. The new simulation model is shown in Figure 6.1, where 41CD30 is represented by its steel core only and the iron-to-iron distance between the two magnets is 21.4 cm. After performing three-dimensional multipole expansion of the simulation data, the change in the linear focusing function of 30Q58 due to the presence of 41CD30 is obtained from this model and the 30Q58 model in Figure 5.1. The effect on beam optics due to magnetic interference between 30Q58 and 41CD30 can be computed as a first-order perturbation. The linear transfer matrices are obtained and correct hard edge models are constructed for a perturbed quad. The results are also compared with that from a conventional hard edge model.

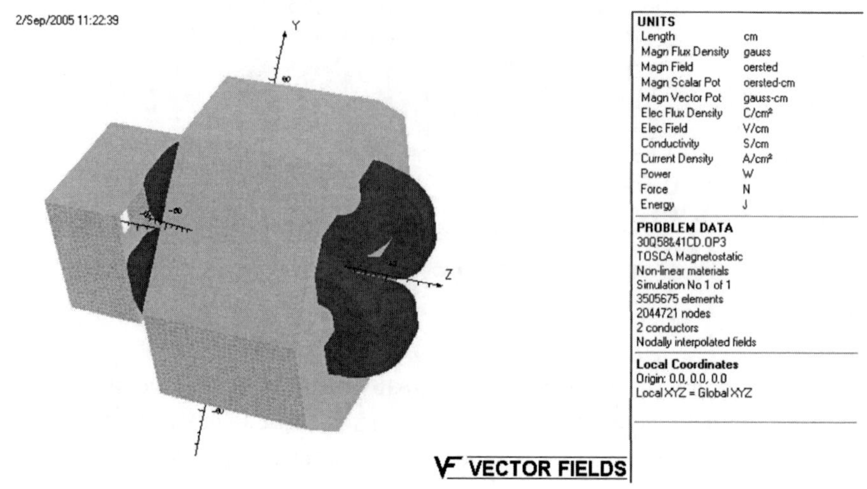

Figure 6.1. Simulation model for 30Q58 and 41CD30.

6.1. Change in Linear Focusing Function

When a magnetic corrector is present in the vicinity of a quadrupole, the quadrupole field distribution is changed. In general, the integrated gradient of the quadrupole is reduced since the quadrupole flux lines are terminated by the corrector iron core. To further understand the effect on particle optics due to the magnetic interference, we first need to compute the linear focusing function of the model of 30Q58 plus 41CD30 through the three-dimensional multipole expansion. The result is shown in Figure 6.2, where k(z) is the linear focusing function of 30Q58 plus 41CD30, and $k_1(z)$ is the linear focusing function for 30Q58 alone. The two linear focusing functions with and without 41CD30 overlap very well in most of the region except around z = -0.5 m, where the corrector is located. This can be seen much more clearly in Figure 6.3, where we plot the difference, $\delta k(z) = k(z)-k_1(z)$, in the linear focusing functions of 30Q58 with and without 41CD30. It is the change in this linear focusing function that determines the extent of the interference, which can be precisely computed.

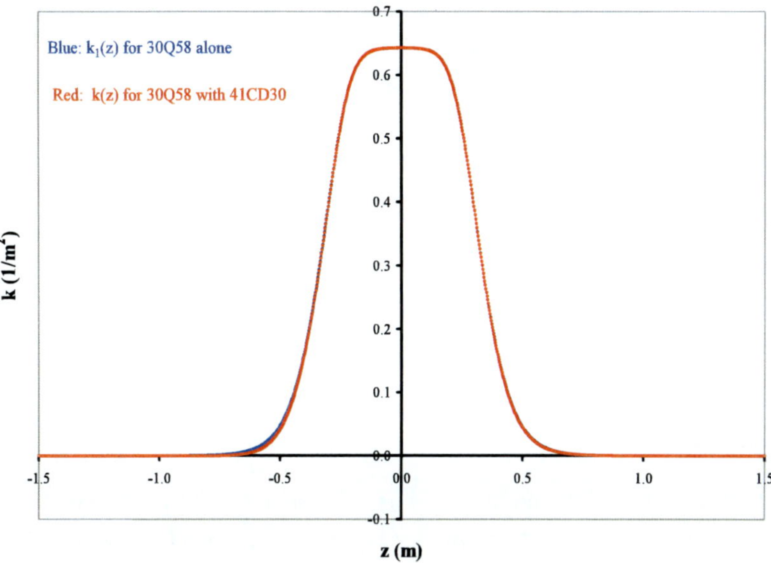

Figure 6.2. Linear focusing functions for 30Q58 with and without 41CD30.

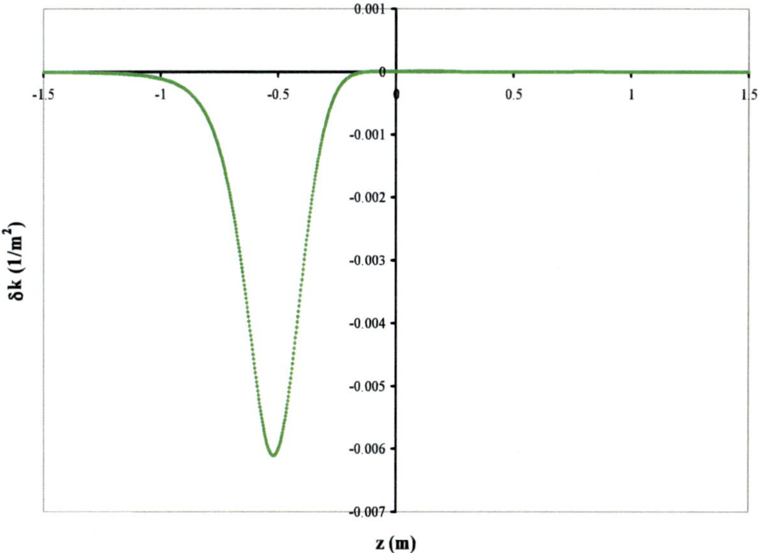

Figure 6.3. Linear focusing function difference for 30Q58 with and without 41CD30.

6.2. Magnetic Interference as a First-Order Perturbation

The integration of $\delta k(z)$ in Figure 6.3 from z_1 = -1.3 m to z_2 = 1.3 m is less than 1% of the $k_1(z)$ integration over the same region. Thus, the magnetic interference between 30Q58 and 41CD30 can be treated as a first-order perturbation [6, 7]. The equations of motion for δx and δy, which are the changes in particle trajectories due to the perturbation, are thus expressed by

$$\delta x'' + k_1(z)\delta x = -\delta k(z)x\,, \tag{6.1a}$$

$$\delta y'' - k_1(z)\delta y = \delta k(z)y\,. \tag{6.1b}$$

Here the linear focusing function $k_1(z)$ is the unperturbed parameter for 30Q58 alone, while $\delta k(z)$ is the perturbation shown in Figure 6.3. The second-order terms $\delta k*\delta x$ and $\delta k*\delta y$ on the right hand side of Eqs. (6.1a, b) are neglected. The established procedure for solving Eqs. (6.1a, b) is to replace x and y on the right hand side by the appropriate unperturbed solutions, i.e. to use $M_{11}(z)$ of the matrices (5.10a) and (5.10b) for the ray with $x_1 = 1$ and $x_1' = 0$; and to use $M_{12}(z)$ for the ray with $x_1 = 0$ and $x_1' = 1$, etc. The numerical solutions for $\delta \mathbf{x}(z)$ and $\delta \mathbf{y}(z)$ are plotted in Figures 6.4 and 6.5, respectively. At the exit z_2 = 1.3 m, the changes in the linear transfer matrix of 30Q58 alone due to the perturbation of 41CD30 are for the x-motion

$$\delta \mathbf{M}_x = \begin{bmatrix} 2.846 & 2.141 \\ 1.403 & 1.076 \end{bmatrix} \bullet 10^{-3}, \tag{6.2a}$$

and for the y-motion

$$\delta \mathbf{M}_y = \begin{bmatrix} -3.888 & -2.892 \\ -2.264 & -1.700 \end{bmatrix} \bullet 10^{-3}. \tag{6.2b}$$

Another approach for the interference problem is to solve Eq. (5.9) directly with the linear focusing function k(z) in Figure 6.2. This is more straightforward and easier to obtain the final results. With this approach, we can directly find the transfer matrix from z_1 = -1.3 m to z_2 = 1.3 m for the x-motion in 30Q58 plus 41CD30 as

$$\mathbf{M}_{x1} = \begin{bmatrix} 0.46391 & 1.91802 \\ -0.40960 & 0.46214 \end{bmatrix}. \tag{6.3a}$$

For the y-motion in 30Q58 plus 41CD30, the linear transfer matrix is

$$\mathbf{M}_{y1} = \begin{bmatrix} 1.58846 & 3.33959 \\ 0.45715 & 1.59065 \end{bmatrix}. \tag{6.3b}$$

It is easy to verify numerically that the matrices in (6.3a) and (6.3b) are the sum of the matrices in (5.10a, b) and (6.2a, b), i.e. $\mathbf{M}_{x(y)1} = \mathbf{M}_{x(y)} + \delta\mathbf{M}_{x(y)}$, as we expected. Thus, we have two equivalent approaches to accurately calculate the effect of the magnetic interference from the linear focusing functions: one is to find the perturbations only first; and another is to yield the net results.

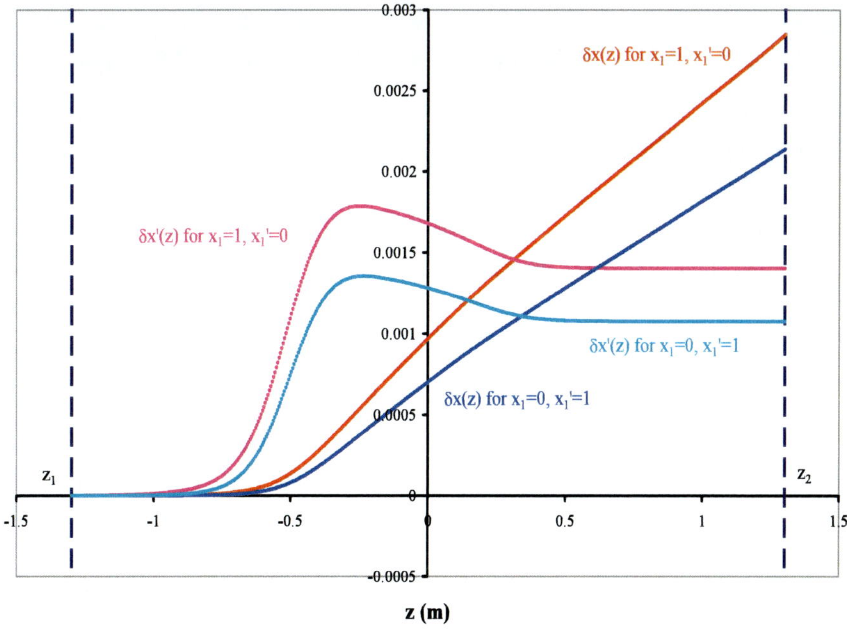

Figure 6.4. Perturbations on particle trajectories for the x-motion in 30Q58 plus 41CD30.

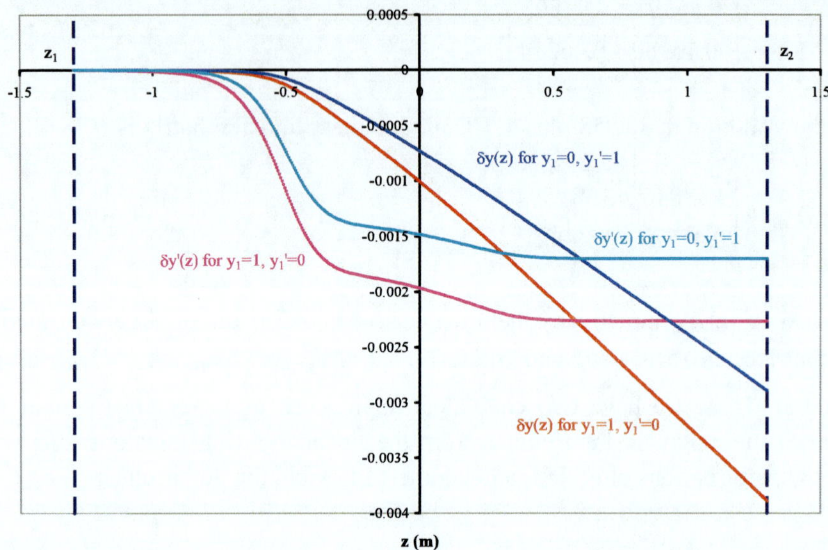

Figure 6.5. Perturbations on particle trajectories for the y-motion in 30Q58 plus 41CD30.

6.3. Hard Edge Models for a Perturbed Quad

In matrices (6.3a) and (6.3b) the two diagonal elements are not equal, and the matrices now have three independent elements due to the asymmetry of the quad perturbed by an adjacent corrector. A symmetric equivalent hard edge model as derived before does not exist any more, since that model has only two independent parameters.

There are a few options to derive hard edge models for an asymmetric quad. The first one is to keep the hard edge models from the first-order matrices for a single 30Q58 unchanged, as shown in Figure 5.11. The perturbation due to 41CD30 can be represented by another perturbation hard edge model starting at z = -L/2-w and ending at z = -L/2. This is to require

$$\mathbf{M}_1 = \mathbf{D}\mathbf{H}\mathbf{H}_1\mathbf{D}_1 . \tag{6.4}$$

Here $\mathbf{M_1}$ is the perturbed matrices (6.3a) and (6.3b), $\mathbf{D}$ and $\mathbf{H}$ are the matrices in Eq. (5.13) for the unperturbed quad, $\mathbf{H}_1$ is the desired perturbation hard edge model due to the interference, and $\mathbf{D}_1$ is a drift space matrix with a distance of $-L/2-w-z_1$. By employing Eq. (5.13), we find the relation

$$\mathbf{D}_1^{-1}\mathbf{H}_1^{-1} = \mathbf{M}_1^{-1}\mathbf{M}\mathbf{D}^{-1} \equiv \mathbf{Q}. \tag{6.5}$$

The $\mathbf{H_1}$ parameters can then be found as

$$\varphi = \kappa w = C^{-1}(Q_{22}), \tag{6.6a}$$

$$\kappa = \mp Q_{21}/S. \tag{6.6b}$$

As before, C and S are cos and sin for focusing, and cosh and sinh for defocusing. The numerical values show that for the x-motion the perturbation hard edge model $\mathbf{H}_1$ has an effective lens strength $\varphi = 1.36^\circ$, height $k_0 = -0.005859\ m^{-2}$, and length $w = 0.3101$ m. As we expected from the polarity of $\delta k(z)$ in Figure 6.3, $\mathbf{H}_1$ represents a defocusing effect for a focusing 30Q58. For the y-motion, the perturbation hard edge model $\mathbf{H}_1$ has an effective lens strength $\varphi = 1.35^\circ$, height $k_0 = 0.00595\ m^{-2}$, and length $w = 0.3058$ m. It is a focusing element prior to the unperturbed hard edge model $\mathbf{H}$. This is shown in Figure 6.6, where the main hard edge models are the same as those in Figure 5.11 for a single 30Q58, and the height of the perturbation hard edge models are magnified by a factor of 10 for clarity. A conventional hard edge model for the perturbed 30Q58 is also plotted for the purpose of comparison. This model is still symmetric with respect to $z = 0$, and its height k_0 and length L are obtained in a conventional way as before.

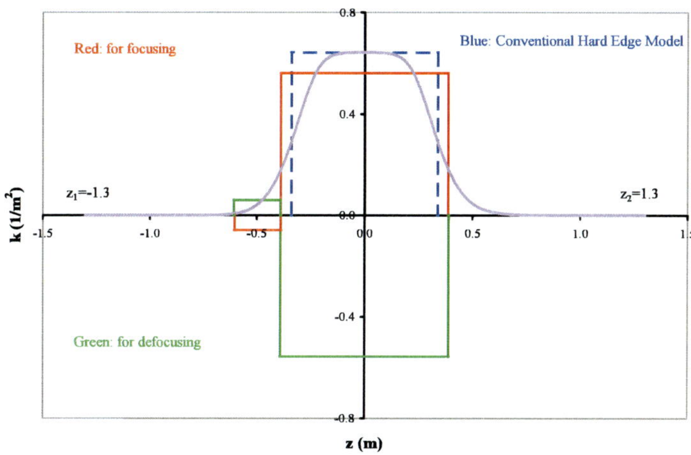

Figure 6.6. Hard edge models for magnetic interference in 30Q58 plus 41CD30.

For the second option, we may consider an equivalent hard edge model that is asymmetric with respect to the magnet center (z = 0) and is sandwiched in between two drift spaces of unequal distances. This is to require

$$\mathbf{M}_1 = \mathbf{D}_2\mathbf{H}\mathbf{D}_1. \tag{6.7a}$$

Here $\mathbf{D}_1$ is the drift matrix from z_1 to $-L_1$

$$\mathbf{D}_1 = \begin{bmatrix} 1 & -L_1 - z_1 \\ 0 & 1 \end{bmatrix}, \tag{6.7b}$$

and $\mathbf{D}_2$ is the drift matrix from L_2 to z_2,

$$\mathbf{D}_2 = \begin{bmatrix} 1 & z_2 - L_2 \\ 0 & 1 \end{bmatrix}, \tag{6.7c}$$

and $\mathbf{H}$ is the transfer matrix for a hard edge quad of length $L = L_1+L_2$, as defined before. It can be shown that the effective lens strength φ of the hard edge model can be found from the following transcendental equation with the matrix elements from the matrices $\mathbf{M}_1$ in (6.3a) and (6.3b)

$$\frac{M_{11}+M_{22}}{2} - M_{21}z_2 = C \pm \frac{1}{2}\varphi S\,. \tag{6.8a}$$

Other hard edge parameters for $\mathbf{H}$ are obtained as

$$L_1 = l - \frac{M_{22}-C}{M_{21}}, \tag{6.8b}$$

$$L_2 = l - \frac{M_{11}-C}{M_{21}}, \tag{6.8c}$$

$$\kappa = \frac{M_{21}}{\mp S}. \tag{6.8d}$$

It is easy to see that if the quad is symmetric with respect to z = 0, i.e. $M_{11} = M_{22}$, we recover the previous hard edge model with $L_1 = L_2 = L/2$.

With the numerical values for the focusing and defocusing matrices, we compute the asymmetric hard edge models for 30Q58 plus 41CD30. The resulting plot would be similar to that in Figure 5.11, but with different parameters as listed in Table 6.1. In comparison with the data in Table 5.2, we see that the perturbation from 41CD30 reduces the 30Q58 total hard edge length L and the effective lens strength φ, and slightly increases its hard edge height k_0. In addition, the interference makes the hard edge model asymmetric with respect to the quad center at z = 0.

Table 6.1. Hard edge models for 38Q58 plus 41CD30

	Conventional	Focusing	Defocusing
$-L_1$ (m)	-0.3370	-0.3815	-0.3829
L_2 (m)	0.3370	0.3858	0.3877
L (m)	0.6740	0.7673	0.7706
k_0 (1/m2)	0.6423	0.5646	0.5615
k_0*L (1/m)	0.4329	0.4332	0.4327
φ (degree)	30.95	33.03	33.09
f_x (m)	2.4261	2.4414	
f_y (m)	-2.2011		-2.1875

REFERENCES

[1] Y. Papaphilippou, Y. Y. Lee, and W. Meng, "Impact of magnetic field interference in the SNS ring," in Proceedings of the 2001 Particle Accelerator Conference, Chicago, IL, June 18–22, 2001, eds. P. Lucas and S. Webber (IEEE Piscataway, NJ, 2001), p. 1667.

[2] J. G. Wang, N. Tsoupas, M. Venturini, "3D simulation studies of SNS ring doublet magnets," in Proceedings of the 2005 Particle Accelerator Conference, Knoxville, TN, May 16–20, 2005, ed. C. Horak (IEEE Catalog Number 05CH37623C, 2005), p. 3865; also see J. G. Wang, SNS-NOTE-MAG-154, October 19, 2005.

[3] D. Raparia and W. Meng, "Magnet analysis and engineering support," in ASAC (Accelerator System Advisory Committee) Review of the SNS Project, September 2004.

[4] A. Gorda, C. Dimopoulou, A. Dolinskii, F. Nolden, and M. Steck, "Field interference of magnets in the large acceptance storage ring of the FAIR project," in Proceedings of the Eleventh European Particle Accelerator Conference, Genoa, Italy, June 23–27, 2008, eds. I. Andrian et al. (EPS – AG, 2008), p. 3113.

[5] J. G. Wang, "Particle optics of quadrupole doublet magnets in Spallation Neutron Source accumulator ring," *Phys. Rev. ST Accel. Beams* 9, 122401 (2006). (http://prst-ab.aps.org/abstract/PRSTAB/v9/i12/e122401); also see J. G. Wang, SNS-NOTE-MAG-168, July 12, 2006.

[6] P. M. Morse and H. Feshbach, *Methods of Theoretical Physics*, McGraw-Hill, New York, 1953.

[7] K. Halbach, "First order perturbation effects in iron-dominated two-dimensional symmetrical multipoles," *Nucl. Instrum. Methods* 74 147 (1969).

Chapter 7

PARTICLE OPTICS IN A QUAD DOUBLET ASSEMBLY

The quadrupole doublet assembly in the SNS accumulator ring consists of two magnetic quadrupoles (30Q58 and 30Q44) and a magnetic dipole corrector (41CD30). The quad 30Q58 acts as a focusing element in the horizontal direction for protons, while 30Q44 is for defocusing horizontally. A three-dimensional simulation model of the quadrupole doublet assembly, as shown in Figure 7.1, has been built to study the particle optics in the devices. The two quads in the assembly have an iron-to-iron distance of 51.4 cm. The distance between 30Q58 and 41CD30 is 21.4 cm, the same as in Figure 6.1. The two quadrupoles 30Q58 and 30Q44 interfere with each other, in addition to the interference between 30Q58 and 41CD30.

We first present the two-dimensional field parameters based on our models, which show the relative strength of the interference. The effect on particle optics can be analyzed in the same procedure as described in Chapter 6. The three-dimensional multipole expansion of the field data from the simulation model is performed first, and the linear focusing function is then computed. The changes in particle motion due to the interference can be obtained from the perturbed ray traces. We construct the hard edge models for the assembly, in which the centers and identity of the two quads remain. The results are compared with a conventional hard edge model. The third-order aberrations in the assembly due to magnetic fringe fields and interference are analyzed and illustrated. The validity of the technique developed here is demonstrated by the agreement between the particle optics computations of the ray traces and the OPERA-3D TRACK command calculations of particle trajectories. The material presented in this chapter can also be found largely in reference [1].

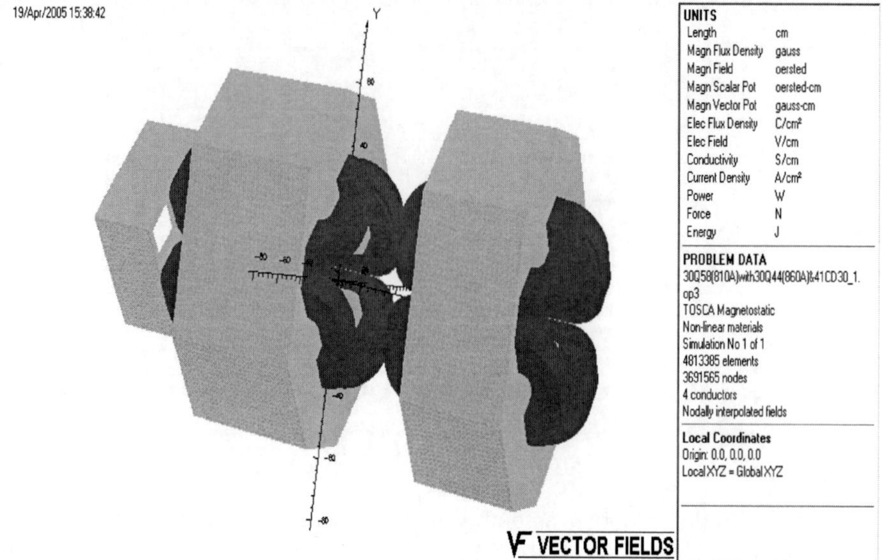

Figure 7.1. Simulation model for SNS ring quadrupole doublet assembly (from left to right: 41CD30 iron core, 30Q58, 30Q44).

7.1. Two-Dimensional Field Parameters

In post-processing of OPERA-3D/TOSCA simulations, we first obtain a two-dimensional field representation for the models as described in Sections 2.2 and 2.6. For quadrupoles, the integrated gradient (G*L) and the integrated harmonics (C_m*L for m > 2) are calculated and expressed by the normalized coefficients in "units".

The results from the different simulation models are listed in Table 7.1. In calculations, the reference radius R_{ref} is 10 cm. The reduction of the integrated gradient of 30Q58 due to its neighbors can reach more than 1%. This effect is probably significant, considering the accuracy of magnetic field requirements for the SNS ring is at a lower 10^{-4} level. The dynamic effect of the interference, i.e. the effect on 30Q58 due to the current variation in 30Q44, appears negligible. The harmonic contents also change, but by less than 2 units. The two-dimensional field parameters indicate the strength of magnetic interference. However, they can not produce correct particle optics in the quad doublet assembly.

Table 7.1 Integrated gradients and harmonics in various models

		G*L (T)	ΔG*L (%)	NC_6 (Unit)	NC_{10} (Unit)	NC_{14} (Unit)
30Q58 (810A)	Alone	2.4576	100%	2.43	7.62	0.94
	w/ 41CD30	2.4488	99.64%	2.63	7.42	0.94
	w/ 30Q44 (860A)	2.4427	99.40%	3.83	7.55	0.95
	w/ 30Q44 (960A)	2.4421	99.37%	4.07	7.62	0.99
	w/ 30Q44 (860A) and 41CD30	2.4325	98.98%	3.25	7.76	0.99
	w/ 30Q44 (960A) and 41CD30	2.4319	98.95%	3.39	7.62	0.98
30Q44 (860A)	Alone	2.0713	100%	1.97	8.01	0.90
	w/ 30Q58 (810A)	2.0548	99.21%	3.01	7.99	0.91
	w/ 30Q58 (810A) and 41CD30	2.0545	99.19%	3.02	8.02	0.92

7.2. Magnetic Fringe and Interference

Following the procedure described in Chapter 4, we obtain magnetic field distributions in the simulation model shown in Figure 7.1, and we further expand the field data into three-dimensional multipoles. From the generalized gradient for m = 2, we can derive the linear focusing function in the quadrupole doublet assembly model according to Eq. (5.5). In Figure 7.2 we plot three linear focusing functions: the blue curve $k_1(z)$ represents the linear focusing function of 30Q58 alone, which is the same curve as in Figure 5.7; the green curve $k_2(z)$ shows the linear focusing function of 30Q44 itself, which is the same curve as in Figure 5.14 but is offset by z = 1.02489 m (this is the distance between the two quad centers); the red curve k(z) is the linear focusing function when the two quads and the corrector steel are combined together in simulation, i.e., in the quadrupole doublet model in Figure 7.1.

The differences between k(z) and $k_1(z)+k_2(z)$ appear around z = 0.5 m between the two quads, as well as around z = -0.5 m due to magnetic interference from corrector 41CD30. A better picture for this effect is shown in Figure 7.3, where we plot the difference between k(z) and $k_1(z)+k_2(z)$, i.e. $\delta k(z) = k(z)-[k_1(z)+k_2(z)]$. It is obvious that the magnetic interference is dominated by the effect of 41CD30 on 30Q58.

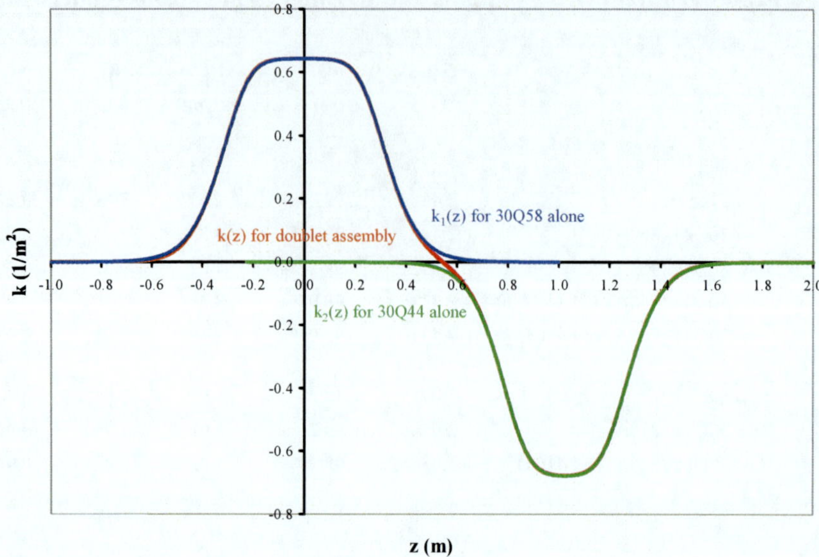

Figure 7.2. Linear focusing functions in doublet assembly.

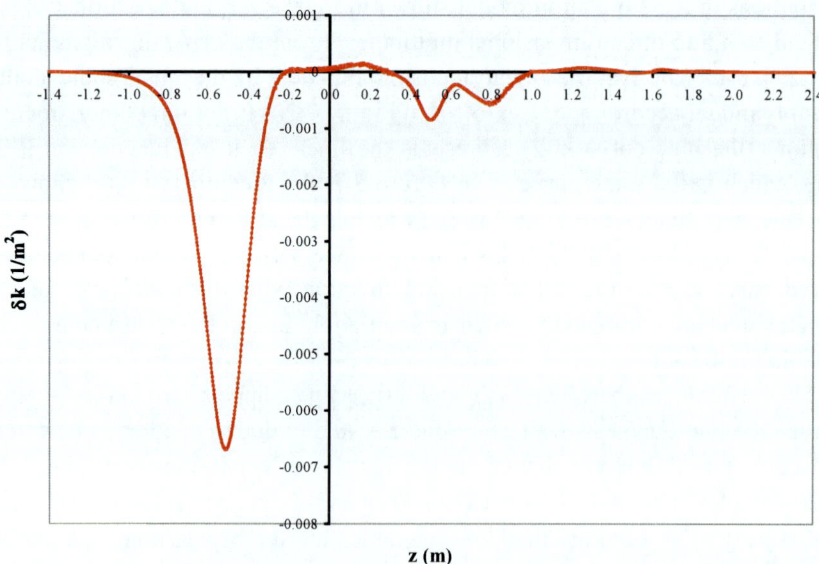

Figure 7.3. Difference between k(z) and $k_1(z)+k_2(z)$.

The analysis of magnetic interference in the quadrupole doublet assembly follows the same procedures as described in Chapter 6. We first use the linearly superimposed focusing function $k_1(z)+k_2(z)$ in Eq. (5.9) to obtain the unperturbed transfer matrices, which do not take into account the effect of magnetic interference. Then, by employing $\delta k(z)$ in Figure 7.3, we can solve Eqs. (6.1a, b) for the perturbations on particle trajectories. These perturbations in the quad doublet assembly are plotted in Figures 7.4 and 7.5. We evaluate these perturbations on the trajectories at an exit point $z_2 = 2.275$ m. This results in the perturbation matrices from $z_1 = -1.3$ m to $z_2 = 2.275$ m as

$$\delta\mathbf{M}_x = \begin{bmatrix} 5.729 & 4.633 \\ 2.536 & 2.080 \end{bmatrix} \bullet 10^{-3}, \qquad (7.1a)$$

$$\delta\mathbf{M}_y = \begin{bmatrix} -5.118 & -4.136 \\ -1.290 & -1.105 \end{bmatrix} \bullet 10^{-3}. \qquad (7.1b)$$

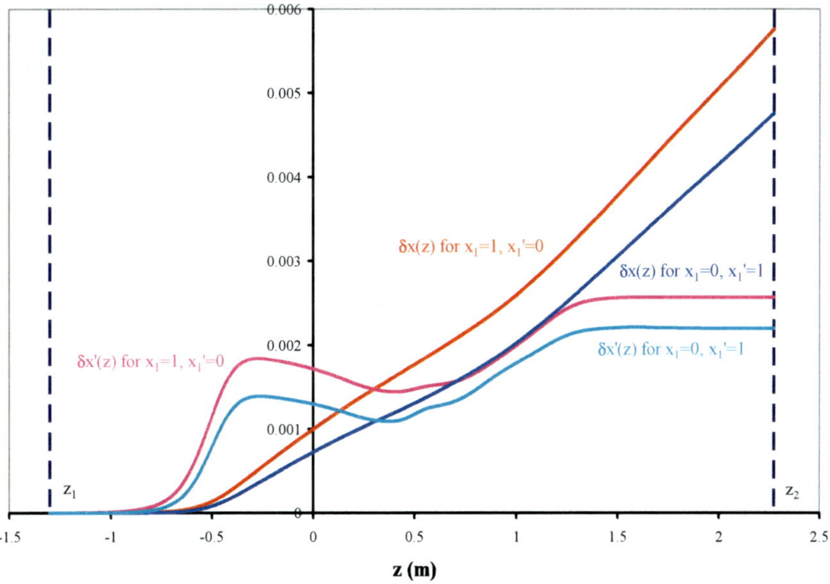

Figure 7.4. Perturbations on the x-motion due to magnetic interference in the quad doublet assembly.

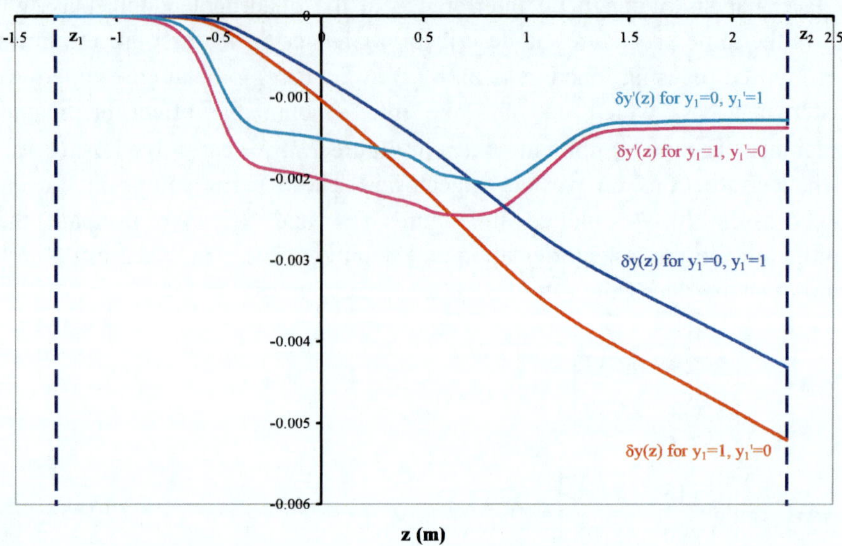

Figure 7.5. Perturbations on the y-motion due to magnetic interference in the quad doublet assembly.

As in the case of 30Q58 plus 41CD30, the sum of those matrices obtained from the unperturbed focusing function $k_1(z)+k_2(z)$ and the perturbation matrices (7.1a, b) leads to the actual transfer matrices, which include the effect of magnetic interference in the assembly. This has been verified numerically. A more straightforward and more efficient way to evaluate the effect of magnetic interference is to directly use the perturbed linear focusing function k(z) as described below.

7.3. Linear Transfer Matrices and Hard Edge Models

By employing the red curve k(z) in Figure 7.2, we can solve Eq. (5.9) directly to obtain particle trajectories for the entire region of the quadrupole doublet assembly, as shown in Figures 7.6 and 7.7. The linear transfer matrices for mapping particles from $z_1 = -1.3$ m to $z_2 = 2.275$ m are then obtained as

$$\mathbf{M}_x = \begin{bmatrix} 0.34458 & 3.20970 \\ -0.18881 & 1.14338 \end{bmatrix}, \tag{7.2a}$$

$$\mathbf{M}_y = \begin{bmatrix} 1.39186 & 3.62363 \\ -0.058763 & 0.56549 \end{bmatrix}. \tag{7.2b}$$

The eigenvalues of these two matrices are $0.74398 \pm 0.66821i$ and $0.97869 \pm 0.20546i$, which correspond to the matrix phase advance of 41.93° and 11.86°, respectively. The focal length of the transfer matrices is $f_x = 5.30$ m and $f_y = 17.02$ m.

These matrices in principle suffice to represent the quad assembly in the commonly used lattice design codes [2–4]. However, many accelerator designers prefer hard edge models of magnets in lattices because the hard edge parameters can be easily scaled with beam energy and magnet current. In order to derive hard edge models, in which the identity and their geometric centers of the two quads still remain, we separate them at z_0 = 0.54455 m, where the linear focusing function k(z) crosses zero. By employing a similar approach as that for 30Q58 plus 41CD30 in Eqs. (6.7) and (6.8), we obtain the hard edge models for the quad doublet assembly as shown in Figure 7.8. The hard edge parameters are listed in Table 7.2.

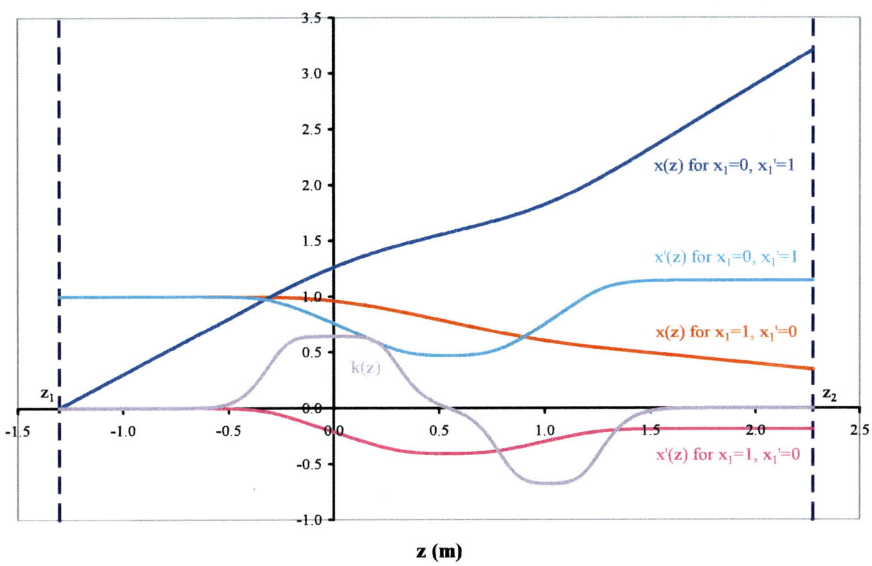

Figure 7.6. Particle trajectories on the x-motion in the quad doublet assembly.

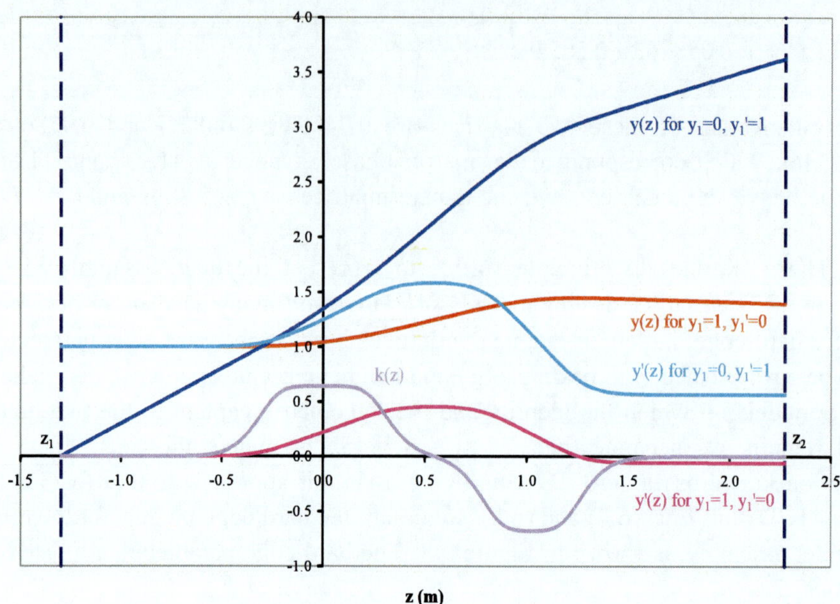

Figure 7.7. Particle trajectories on the y-motion in the quad doublet assembly.

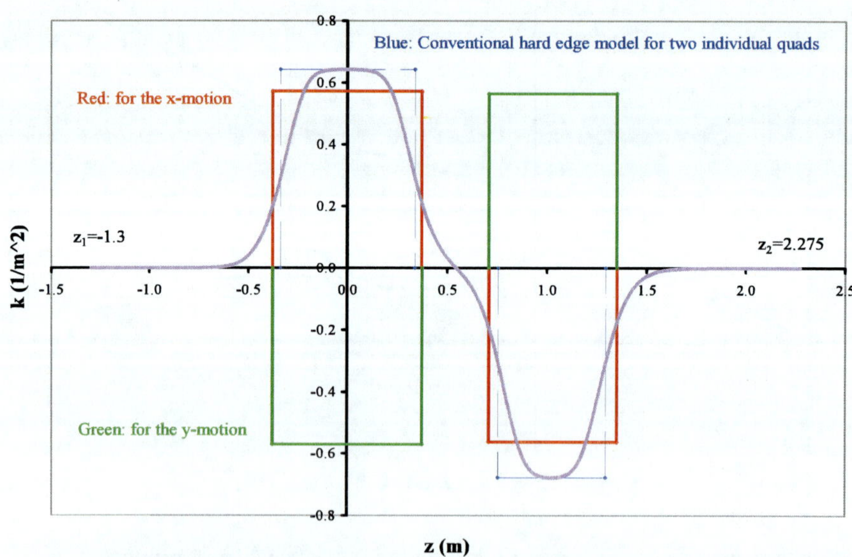

Figure 7.8. Hard edge models for the quad doublet assembly.

Table 7.2. Hard edge models and lens parameters for the quad doublet assembly

	1. First-order theory		2. No interference		3. Conventional	
	x-motion	y-motion	x-motion	y-motion	x-motion	y-motion
			Hard edge parameters			
	30Q58		30Q58		30Q58	
$k_o (1/m^2)$	0.5728	-0.5704	0.5609	-0.5575	0.6423	-0.6423
L_1 (m)	0.3769	0.3785	0.3879	0.3897	0.3385	0.3385
L_2 (m)	0.3726	0.3736	0.3879	0.3897	0.3385	0.3385
L (m)	0.7495	0.7521	0.7757	0.7794	0.6769	0.6769
	30Q44		30Q44		30Q44	
$k_o (1/m^2)$	-0.5624	0.5652	-0.5513	0.5547	-0.6770	0.6770
L_1 (m)	0.3170	0.3161	0.3318	0.3303	0.2704	0.2704
L_2 (m)	0.3274	0.3257	0.3318	0.3303	0.2704	0.2704
L (m)	0.6443	0.6418	0.6635	0.6605	0.5407	0.5407
			Transfer matrix elements from $z_1 = -1.3$ m to $z_2 = 2.275$ m			
M_{11}	0.3446	1.3919	0.3388	1.3971	0.3294	1.3874
M_{12} (m)	3.2097	3.6236	3.2050	3.6278	3.1907	3.6148
M_{21} (1/m)	-0.1888	-0.05876	-0.1914	-0.05736	-0.1966	-0.06275
M_{22}	1.1434	0.5655	1.1411	0.5668	1.1320	0.5573
			Lens parameters			
f(m)	5.30	17.02	5.22	17.43	5.09	15.94

The correct hard edge models for the quad doublet assembly, which are derived rigorously from the first-order linear matrices, have taken into account both the fringe fields and interference. Their correctness can be verified by the fact that these models can produce the same matrices as in (7.2a, b). The numerical errors in calculations between the results from the models and those in (7.2a, b) are shown below:

$$\Delta \mathbf{M}_x = \begin{bmatrix} -4.9 & 1.3 \\ -14 & 0.81 \end{bmatrix} \bullet 10^{-6}, \tag{7.3a}$$

$$\Delta \mathbf{M}_y = \begin{bmatrix} -2.5 & -4.0 \\ 37 & 29 \end{bmatrix} \bullet 10^{-6}. \tag{7.3b}$$

The conventional hard edge model for the quad doublet assembly, as indicated by the blue dotted lines in Figure 7.8, is obtained simply by positioning the conventional hard edge models of the individual 30Q58 and 30Q44 at their lattice locations. The focusing strength and magnetic length in the conventional hard edge models is 0.6423 $1/m^2$ and 0.6769 m for 30Q58 (Table 5.2), and -0.6770 $1/m^2$ and 0.5407 m for 30Q44 (Table 5.4), respectively. This leaves three drift distances between $z_1 = -1.3$ m and $z_2 = 2.275$ m: 0.96155 m before 30Q58, 0.41609 m between the two quads, and 0.97976 m after 30Q44. For the x-motion, 30Q58 is a focusing quad for protons, while 30Q44 is defocusing in the x-direction, and verse versa. Concatenations of the five matrices (two quads and three drifts) yield the conventional hard edge transfer matrices for the quad doublet assembly as

$$\mathbf{M}_x' = \begin{bmatrix} 0.32935 & 3.19067 \\ -0.19656 & 1.13203 \end{bmatrix}, \tag{7.4a}$$

$$\mathbf{M}_y' = \begin{bmatrix} 1.3874 & 3.6148 \\ -0.062753 & 0.55726 \end{bmatrix}. \tag{7.4b}$$

The eigenvalues of these two matrices are $0.73069 \pm 0.68271i$ and $0.97235 \pm 0.23354i$, which correspond to the matrix phase advance of 43.06° and 13.51°, respectively. The focusing length is $f_x = 5.09$ m and $f_y = 15.93$ m.

In comparison with the matrices in (7.2a, b), which are obtained by the first-order theory, the difference is significant. The relative errors for the matrix elements, calculated as $\delta\mathbf{M}_{x/y} = (\mathbf{M}'_{x/y}-\mathbf{M}_{x/y})/\mathbf{M}_{x/y}$, are

$$\delta\mathbf{M}_x = \begin{bmatrix} -4.4\% & -0.59\% \\ 4.1\% & -0.99\% \end{bmatrix}, \tag{7.5a}$$

$$\delta\mathbf{M}_y = \begin{bmatrix} -0.32\% & -0.25\% \\ 6.79\% & -1.46\% \end{bmatrix}. \tag{7.5b}$$

The second hard edge model for the quad doublet assembly in Table 7.2 is obtained by employing the correct hard edge models for 30Q58 and 30Q44 from the first-order theory. This includes the magnetic fringe fields of each quad, but without any magnetic interference. For the x-motion, the focusing strength and magnetic length in the first-order hard edge models is 0.5609 $1/m^2$ and 0.7757 m for 30Q58 (Table 5.2), and -0.5513 $1/m^2$ and 0.6635 m for 30Q44 (Table 5.4), respectively. This leaves three drift distances between $z_1 = -1.3$ m and $z_2 = 2.275$ m: 0.91215 m before 30Q58, 0.30529 m between the two quads, and 0.91836 m after 30Q44. For the y-motion, the focusing strength and magnetic length is -0.5575 $1/m^2$ and 0.7794 m for 30Q58, and they are 0.5547 $1/m^2$ and 0.6605 m for 30Q44. The three drift distances between $z_1 = -1.3$ m and $z_2 = 2.275$ m are: 0.9103 m before 30Q58, 0.30494 m between the two quads, and 0.91986 m after 30Q44. The matrix calculations yield

$$\mathbf{M}_x'' = \begin{bmatrix} 0.338775 & 3.20495 \\ -0.191403 & 1.14106 \end{bmatrix}, \tag{7.6a}$$

$$\mathbf{M}_y'' = \begin{bmatrix} 1.39705 & 3.62783 \\ -0.057356 & 0.566853 \end{bmatrix}. \tag{7.6b}$$

The eigenvalues of these two matrices are $0.739918 \pm 0.672697i$ and $0.981952 \pm 0.18913i$, which correspond to the matrix phase advances of 42.28° and 10.90°, respectively. The focal length is $f_x = 5.22$ m and $f_y = 17.43$ m.

In comparison with the correct linear matrices in (7.2a, b), the relative errors for the matrix elements, calculated as $\delta\mathbf{M}_{x/y} = (\mathbf{M}''_{x/y} - \mathbf{M}_{x/y})/\mathbf{M}_{x/y}$, are shown below:

$$\delta\mathbf{M}_x = \begin{bmatrix} -1.7\% & -0.15\% \\ 1.4\% & -0.20\% \end{bmatrix}, \tag{7.7a}$$

$$\delta\mathbf{M}_y = \begin{bmatrix} 0.37\% & 0.12\% \\ -2.4\% & 0.24\% \end{bmatrix}. \tag{7.7b}$$

These errors are smaller than the errors in (7.5a, b). The matrix M_y'' results in a focal length of $f_y = 17.43$ m, which is closer to 17.02 m than from (7.2b). Its phase advance of 10.90° is also closer to 11.86° than from (7.2b). This better agreement comes from taking into account the fringe field in both individual 30Q58 and 30Q44, but not yet the magnetic interference.

The hard edge parameters, transfer matrix elements, and the lens parameters for the three different models are listed in Table 7.2. In these models, 30Q58 center is at z = 0, while it is at z = 1.02489 m for 30Q44. The two quads are sandwiched by three drift distances starting from $z_1 = -1.3$ m and ending at $z_2 = 2.275$ m. It can be seen that the discrepancies between the conventional hard edge model and the correct first-order model are quite significant. The focusing lengths f_x and f_y in the first-order model are 5.30 and 17.02 m, respectively, while they are 5.09 and 15.93 m in the conventional model. The difference in the matrix phase advances between (7.2a, b) and (7.4a, b) is also rather large. The parameters in the second hard edge model are closer to the correct ones than the conventional hard edge model, since the magnetic fringe fields are included.

It should be noted that the correct first-order hard edge models of 30Q58 and 30Q44 in the quadrupole doublet assembly depend on the given magnet currents. This is due to the overlapping magnetic fringe fields of the two quadrupoles. In order to keep the centers and identity of the two quads unchanged in the hard edge models, we tried to separate the magnetic fields of the two quads at the longitudinal position z_0, where the linear focusing function k(z) crosses zero. Nevertheless, this longitudinal position z_0 changes with the magnet currents. Thus, it would be difficult to use these hard edge models in some applications, such as the optimization of the quadrupole settings for prescribed lattice parameters.

7.4. THIRD-ORDER ABERRATIONS

The third-order aberrations in the SNS quadrupole doublet assembly can be calculated in the same way as in Section 5.7. Here we employ Eqs. (5.18a) and (5.18b) to compute the third-order aberrations [$\Delta x(z)$, $\Delta x'(z)$] on the x-plane due to the aberration coefficients P_0 and P_3. The corresponding two first-order rays are the one with $x_1 = 0.1$ m and $x_1' = 0$, and the other with $x_1 = 0$ and $x_1' = 0.05$, respectively. The results are plotted in Figure 7.9. At $z_2 = 2.275$ m, $\Delta x = -2.10\text{x}10^{-04}$ m, $\Delta x' = -1.10\text{x}10^{-04}$ due to P_0, and $\Delta x = -9.83\text{x}10^{-05}$ m, $\Delta x' = -5.81\text{x}10^{-05}$ due to P_3. They represent the relative changes of -0.61%, -0.58%, -0.061%, and -0.10%, respectively, in comparison with the first-order matrix solutions which are the elements in $\mathbf{M}_x$ in (7.2a, b) scaled by the initial parameters. Similarly, we use Eqs. (5.19a) and (5.19b) to calculate the third-order aberrations on the y-plane due to the aberration coefficients P_0 and P_3. The results are plotted in Figure 7.10. At $z_2 = 2.275$ m, $\Delta y = -3.84\text{x}10^{-04}$ m, $\Delta y' = -2.47\text{x}10^{-04}$ due to P_0, and $\Delta y = -3.22\text{x}10^{-04}$ m, and $\Delta y' = -2.39\text{x}10^{-04}$ due to P_3. They represent the relative changes of -0.28%, 4.20%, -0.18%, and -0.85%, respectively, from the first-order results.

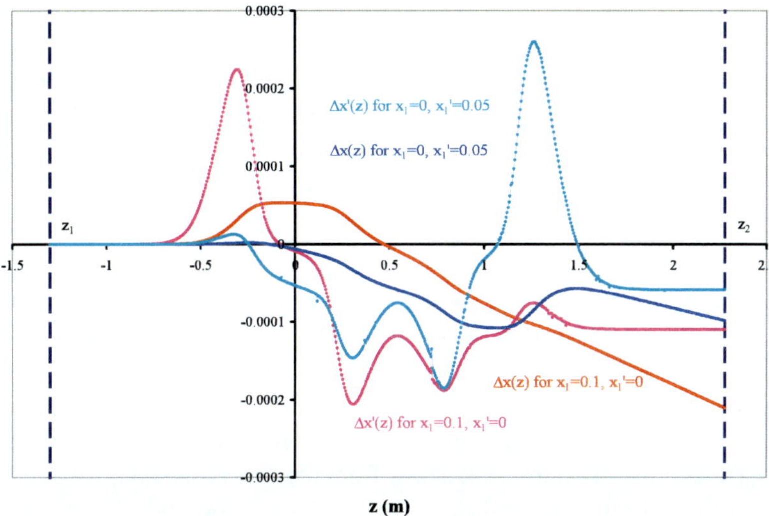

Figure 7.9. Third-order aberrations on the x-plane through the quadrupole doublet assembly.

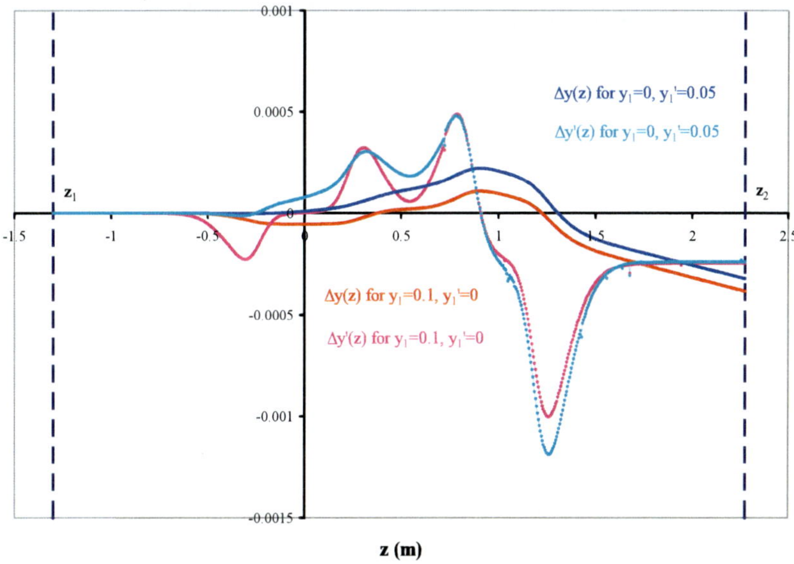

Figure 7.10. Third-order aberrations on the y-plane through the quadrupole doublet assembly.

7.5. Verification of Particle Trajectories

OPERA-3D has a built-in TRACK command for calculating trajectories of charged particles by directly integrating the equations of motion in simulated magnetic fields [5]. We use this command to compute the particle trajectories in the quad doublet assembly model shown in Figure 7.1. We launch a proton particle of 1 GeV in energy at $z_1 = -1.3$ m with different initial conditions: two rays on the x-plane with $x_1 = 0.1$ m, $x_1' = 0$ and $x_1 = 0$, $x_1' = 0.05$; two rays on the y-plane with $y_1 = 0.1$ m, $y_1' = 0$ and $y_1 = 0$, $y_1' = 0.05$. The code calculates the proton particle trajectories through the dipole corrector and the quadrupole doublet until the end of the assembly at $z_2 = 2.275$ m. We plot the trajectories on the x-plane from the OPERA-3D TRACK command in Figure 7.11 and the trajectories on the y-plane in Figure 7.12. In order to compare these particle trajectories with the ray traces calculated from the matrix method, we overlay the ray traces from the first-order calculation in Figures 7.6 and 7.7 scaled by the initial parameters, as shown in Figures 7.11 and 7.12 by green and orange color. The agreement between the OPERA-3D TRACK calculation and the matrix method is excellent. Since the two methods are independent, this proves that the computation of particle optics with magnetic fringe fields and interference developed above can be trusted.

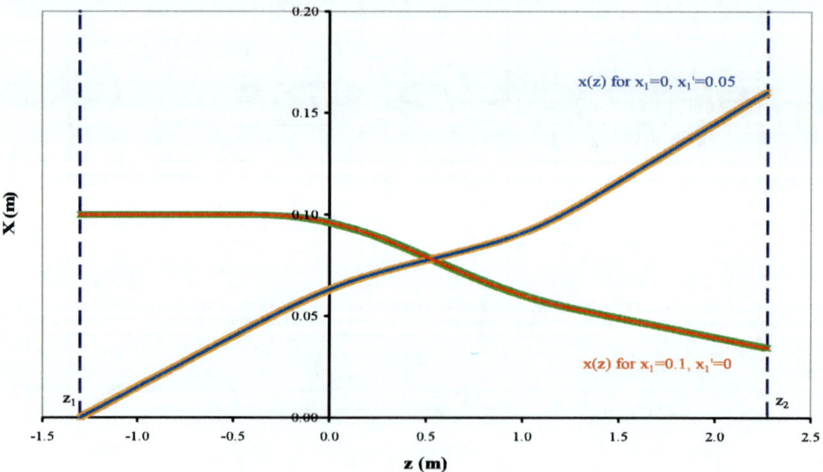

Figure 7.11. Ray traces for the x-motion from OPERA-3D (blue and red) and matrix method.

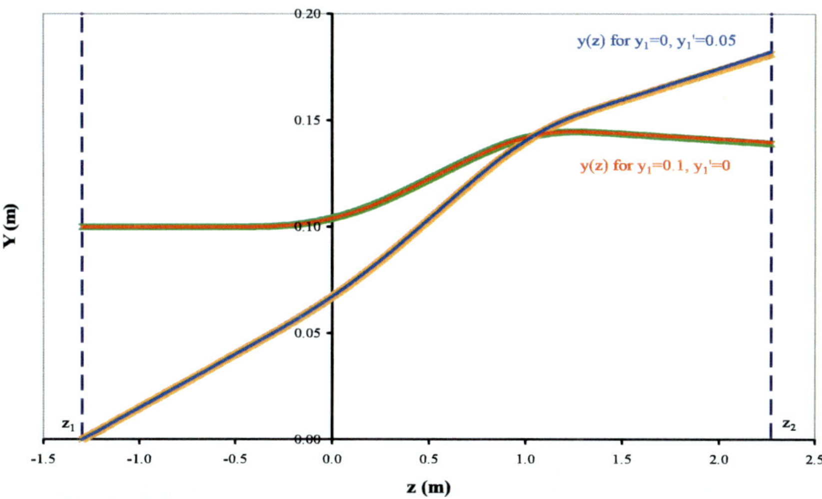

Figure 7.12. Ray traces for the y-motion from OPERA-3D (blue and red) and matrix method.

REFERENCES

[1] J. G. Wang, "Particle optics of quadrupole doublet magnets in Spallation Neutron Source accumulator ring," *Phys. Rev. ST Accel. Beams* 9, 122401 (2006). (http://prst-ab.aps.org/abstract/PRSTAB/v9/i12/e122401); also see J. G. Wang, SNS-NOTE-MAG-168, July 12, 2006.

[2] J. A. Holmes, S. Cousineau, V. V. Danilov, S. Henderson, A. Shishlo, Y. Sato, W. Chou, L. Michelotti, and F. Ostiguy, "ORBIT: Beam dynamics calculations for high intensity rings," *in The ICFA Beam Dynamics Newsletter*, **30**, 2003.

[3] D. C. Carey, K. L. Brown, and F. Rothacker, *Third-Order TRANSPORT with MAD Input, A Computer Program for Designing Charged Particle Beam Transport Systems,* SLAC-R-530, Fermilab-pub-98-310, UC-414, 1998.

[4] H. Grote and F. C. Iselin, *The MAD Program (Methodical Accelerator Design),* CERN/SL/90-13 (AP), 1990.

[5] OPERA-3D User Guide and Reference Manual, Vector Fields Software of Cobham Technical Services, Oxford, England; also see http://www.vectorfields.com/

Chapter 8

PARTICLE TRACKING IN BEAM LINES

The particle optics in magnetic quadrupoles presented in Chapters 5–7 reflects realistic field distributions including magnetic fringe fields and their interference in the devices. The transfer matrices and lens parameters thus generated can be employed in the beam dynamics codes such as TRANSPORT, MAD, and ORBIT for particle tracking in beam transfer lines and rings. However, this approach is not always convenient or efficient in practice due to the rather complex procedures of expanding the fields to three-dimensional multipoles, solving the linear trajectory equations, generating the transfer matrices, and computing the equivalent hard edge models. For short beam lines involving not many magnetic devices, a better approach is to directly compute particle trajectories in beam line magnet assemblies through accurate three-dimensional magnet modeling. The example presented in Section 7.5 illustrates this method.

Another example of particle tracking in the SNS ring injection dump beam line is presented in this chapter. The subject is associated with the problems of heavy particle losses in the injection dump beam line during the SNS ring commissioning and early operation. We first give a brief introduction to the SNS ring injection where quite a few magnets with large apertures are densely packed. Various injection constraints in the system are described, and these dictate the magnet settings in the system. They are very crucial to the waste beam transport, as well as to obtaining good injection and a good equilibrium orbit (closed orbit) for the circulating proton beam. Three-dimensional simulation models consisting of three injection chicane dipoles and an Injection Dump Septum Magnet (IDSM) are built. The magnet performance and field distributions are briefly described, and the initial conditions for particle tracking are discussed. Three-dimensional particle trajectories through the dipoles and IDSM show clearly where and why the particles are lost. Further studies lead to the remedies to the problems.

8.1. SNS Ring Injection and Waste Beam Losses

The SNS accelerator employs non-Liouvillian injection [1, 2] to obtain high intensity proton beams in its accumulator ring. As shown in Figure 8.1, a 1 GeV H^- beam from a linac is focused onto a primary foil (F1) in the accumulator ring to produce a proton beam through charge exchange. A small portion of the H^- particles is not fully stripped by F1 and becomes H^0 particles. The probability that the injected H^- particles pass through the foil and remain as H^- particles is negligibly small, but some incoming H^- particles may miss the foil. The H^0 and H^- particles after F1 propagate forward through chicane dipole D3 and strike a secondary foil (F2), after which they all should be fully stripped to the proton particles. These particles are the so-called waste beams that are bent by a chicane dipole D4 and are then transported through the IDSM. The IDSM is a magnetic dipole with significant gradient component, as explained in Subsection 8.3.2. It not only bends the two waste beams further away from the circulating proton beam line, but also provides horizontal focusing for each of the waste beams. After the IDSM the waste beams go through a horizontally-defocusing quadrupole magnet and are then transported to an injection dump (IDump). The waste beams are designated as H^--proton beam and H^0-proton beam according to their origin. The whole system behaves like a particle spectrometer for charge separation.

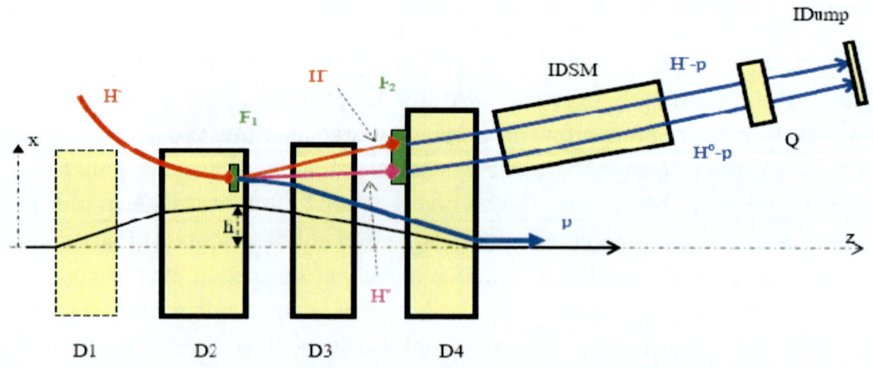

Figure 8.1. Injection dump beam line schematic.

During the SNS ring commissioning and early operation, the entire injection dump beam line, especially the IDSM, suffered high beam losses. After exhausting all of the operational means, the beam losses still remained, and this limited the ramp up of the SNS ring beam power. In order to understand why and where the waste beam particles were lost, we performed three-dimensional modeling of the injection dump beam line consisting of three chicane dipoles and an IDSM. The models took automatically into account magnetic fringe fields and interference among these magnets. The particle trajectories in the models were calculated without complicated particle optics analyses of each device. The studies clearly showed some design and operation problems that caused the waste beam losses in the injection dump beam line. These included incorrect chicane dipole settings, incorrect position of a chicane dipole, overly small aperture of the injection dump septum, and inadequate focusing downstream. The studies also provided the remedies to the injection waste beam loss problems.

8.2. INJECTION CONSTRAINTS

The SNS accumulator ring was designed and fabricated at Brookhaven National Lab [3]. Its injection system [4, 5] is integrated into one of four straight sections of the ring. As shown in Figure 8.1, four chicane dipoles D1 - D4 create a static orbit bump. Four kickers (not shown in the figure) in each of the transverse dimensions generate dynamic bumps for painting particles in both the x and y directions. The chicane dipole settings are critical for obtaining good injection and a good equilibrium orbit of the circulating proton beam, as well as for good transport of the waste beams.

Table 8.1 Chicane dipole bending angles (mrad)

Chicane dipole setting	D1	D2	D3	D4
Design [4, 6]	42.0	-46.2	-42.0	46.2
Delivered [7, 8]	42.0	-53.1	-35.6	46.2
Production [9]	42.0	-53.1	-28.3	39.4

As shown in Table 8.1, there existed mainly three chicane dipole settings for the SNS ring. The design setting [4, 6] and the delivered setting [7, 8] were developed during early stages of the project, and they overlooked some injection constraints as described below. The production setting [9] was worked out during the ring commissioning to yield good injection and a good equilibrium orbit for the circulating proton beam, and was employed in early operations. We studied all the three settings in our three-dimensional models for the waste beam transport.

8.2.1. Closed Orbit Bump and Good Injection

The most important condition for setting the four injection chicane dipoles is to make local closure of the injection orbit bump. This results in a good equilibrium orbit for the circulating proton beam. A proton entering D1 on the z-axis of the straight section, passing through the orbit bump, should exit from D4 on the z-axis. Intuitively, the first condition is that the algebraic sum of the four dipole bending angles equals zero, i.e.

$$\theta_1 + \theta_2 + \theta_3 + \theta_4 = 0 . \tag{8.1}$$

A dipole bending angle is measured by the rotation angle between the incident particle trajectory and the exiting particle trajectory before and after the dipole. Our convention is that the bending angles are positive when the proton is bent counter-clockwise horizontally if one looks down from above. Here, θ_1 and θ_4 are positive, while θ_2 and θ_3 are negative. The condition (8.1) guarantees that the proton goes through the orbit bump without gaining any transverse velocity in the x-direction. In addition, the proton also should not develop any offset from the z-axis at the D4 exit. This can be accomplished if, to the first-order approximation, the following condition is satisfied:

$$(d_{12} + d_{23} + d_{34})\theta_1 + (d_{23} + d_{34})\theta_2 + d_{34}\theta_3 = 0 . \tag{8.2}$$

Here d_{12} is the distance between D1 and D2 centers, etc.

To achieve good injection and painting, the incoming H^- beam and a part of the circulating proton beam should merge at the primary foil with zero relative angles. If the injection takes place at a position where the incoming H^- beam is parallel with the z-axis, as designed, the primary foil should be located at the maximum orbit bump. Then, good injection can be achieved by the following condition:

$$\theta_2 = -(1 + \frac{1}{R})\theta_1 , \tag{8.3}$$

where R is the ratio of the D2 field integrals upstream and downstream of the primary foil, and is determined by the D2 field distribution. The static orbit bump amplitude is another important parameter for injection and painting of injected particles in phase space. In this case, the orbit bump amplitude h can be found, for small angle approximation, as

$$h = (d_{12} - \frac{d_{2F}}{R})\theta_1, \tag{8.4}$$

where d_{2F} is the distance between the D2 center and the primary foil.

Given all the mechanical positions, the orbit bump height h, and the ratio R of D2, the four conditions above uniquely determine the four bending angles. In the SNS ring injection section, $d_{12} = 238.1$, $d_{23} = 181.4$, $d_{34} = 202.9$, $d_{2F} = 30.71$ cm, and the D2 field integral ratio is measured to be $R = 3.784$. If we choose the D1 bending angle $\theta_1 = 42.0$ mrad as from the BNL/SNS data, condition (8.4) leads to the orbit bump amplitude $h = 9.66$ cm. And conditions (8.3), (8.2), and (8.1) yield the dipole bending angles $\theta_2 = -53.1$, $\theta_3 = -28.3$, and $\theta_4 = 39.4$ mrad, respectively. This is exactly the same as the production setting in Table 8.1. We can also see that the delivered setting does not satisfy condition (8.1), leading to an unclosed orbit bump with about 1.51 cm offset at the D4 exit.

Another important consideration for the H^- beam injection is the collection of stripped electrons. The primary foil is located in the D2 downstream fringe region, where the dipole field B_y is about 2.5 kG and the field angle (Atan($-B_z/B_y$)) is more than 200 mrad. Thus, the stripped electrons, leaving the primary foil at the speed of the incoming H^- particles, rotate with a Larmor radius of more than 1.2 cm. They should clear the lower edge of the foil on their first turn, and spiral down to a water-cooled carbon collector below [6]. The design setting in Table 8.1 requires a D2 field integral ratio $R = 10$ to satisfy Eq. (8.3). Therefore, the primary foil should be moved further away downstream to z = 47.3 cm instead of 30.71 cm. At this position the dipole field B_y in the as-built magnet with a bending angle of 46.2 mrad is only about 1.34 kG and B_z is -0.274 kG. The stripped electrons would be trapped in the magnetic field. Thus, the design setting is not applicable. This problem was discovered after the magnets were made and measured. The delivered setting was introduced then, but unfortunately, it does not satisfy the closed orbit bump requirement. This leads to large displacement and unstable oscillation about the displaced equilibrium orbit of the circulating proton beam, and thus, to high beam losses in the ring.

8.2.2. Transport of Waste Beams Through IDSM

Survey data [10] show that the downstream quadrupole magnet axis and the z-axis in the ring straight section form an angle of $\theta_T = 12.57^\circ$ (219.4 mrad). For a reference H^0-proton particle, which is initially in parallel with the z-axis and is bent by D4 first and then by the IDSM, the following condition should be satisfied:

$$\theta_T \le \theta_4 + \theta_{s0} < \theta_T + \delta\theta\,. \tag{8.5a}$$

Here θ_{s0} is the IDSM bending angle for the H^0-proton particle, which will be explained in more detail later. And $\delta\theta$ is a positive, small open angle from the downstream quadrupole magnet entrance to the H^0-proton exit at the IDSM. The distance between the IDSM exit face and the quadrupole magnet center is about 6.25 m. Suppose that the reference particle exits the IDSM at a transverse distance of 8 cm from its axis, we would have a roughly estimated $\delta\theta$ ~ 13 mrad. The un-equal sign is used in Eq. (8.5a) because it is desired that the reference H^0-proton particle converges toward the axis of the downstream quadrupole magnet. With the production setting of the chicane dipoles, D4 would bend the H^0-proton particles by 39.4 mrad. An additional bending angle of a little more than 180.0 mrad from the IDSM would be needed for a reference H^0-proton particle.

By the same token, an H^--proton particle after F1 is first bent by the fields between F1 and F2 plus the D4 field, and then by the IDSM. This leads to the following condition:

$$\theta_T - \delta\theta < -\frac{\theta_2}{1+R} - \theta_3 + \theta_4 + \theta_{s-} \le \theta_T\,, \tag{8.5b}$$

where θ_{s-} is the IDSM bending angle for a reference H^--proton particle. Note that $\theta_{s-} \neq \theta_{s0}$ because the IDSM is a gradient dipole magnet as described later. With the production setting for the chicane dipoles, the fields between F1 and F2 yield a bending angle of about 38.9 mrad. An additional bending angle of a little less than 141.1 mrad from the IDSM would be needed for a reference H^--proton particle. Although we use the same notation of the chicane bending angles in Eqs. (8.5a) and (8.5b) as that in Eqs. (8.1) – (8.4), their numerical values may be slightly different due to different paths considered.

As described before, the bending angles of the four chicane dipoles are uniquely determined by Eqs. (8.1) - (8.4). Thus, Eqs (8.5a) and (8.5b) would have to be fulfilled by the IDSM bending angles θ_{s0} and θ_{s-}. But, it is not trivial to satisfy the two equations simultaneously by one septum magnet. It requires certain trajectory positions in the device, which are determined by the initial waste beam positions and angles at the IDSM entrance. In most cases, the two waste beams incident to the IDSM entrance face are not at normal direction to the IDSM face. This is especially true for the H^--proton particles, which are always incident to the IDSM entrance with large face-rotation angles due to additional bending by the fields between F1 and F2. This makes it very difficult to predict their trajectories inside the IDSM without three-dimensional modeling.

8.3. Three-Dimensional Modeling of Injection Waste Beam Dump Line

The design of the SNS ring injection waste beam line was based on two-dimensional calculations, where the particle motion in the y-direction was overlooked, resulting in a very small vertical aperture of the IDSM. In addition, each magnet on the beam line was treated as a conventional hard-edge component, where the effect of magnetic fringe fields and their interference were ignored. This was in principle inaccurate.

We believe that three-dimensional simulation models and three-dimensional particle trajectories are necessary to take into account all three discrete magnetic field components B_y, B_x, and B_z in beam dynamics. The chicane dipoles D2, D3, D4, and the IDSM are all large aperture magnets. They produce significant fringe fields, which affect the particle motion. This effect can be automatically included in three-dimensional analyses. Further, the four magnets are closely packed and their fields overlap. The magnetic interference changes the field distributions in between any two magnets. Thus, good three-dimensional models should include all the four magnets.

8.3.1. Simulation Models

The simulation environment employed in this study is OPERA-3D/TOSCA v.11. The models consist of four magnets, i.e., D2, D3, D4 and IDSM as shown in Figure 8.2, where three particle trajectories are also illustrated. The simulation models are built by the OPERA package "Modeller" instead of the "Pre-

Processor". The Modeller makes it much easier to simulate two or more magnets together.

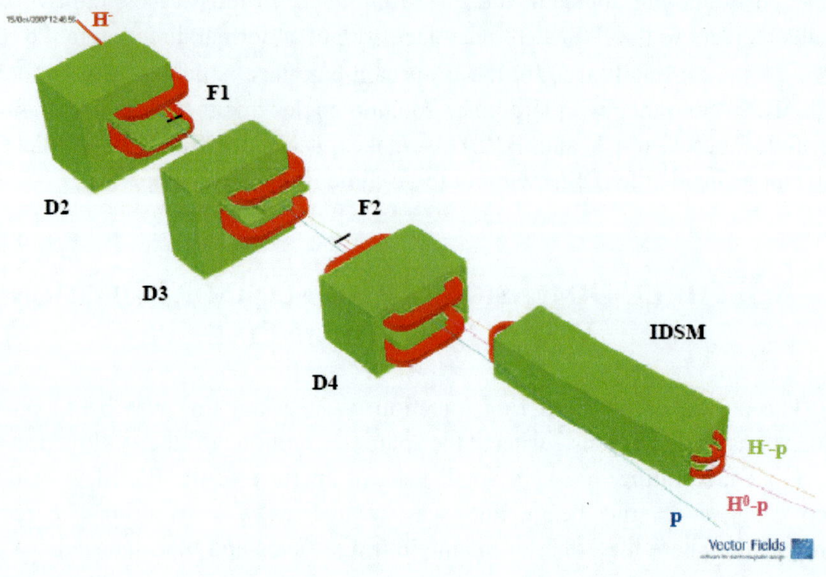

Figure 8.2. Three-dimensional simulation model of SNS ring injection dump beam line.

The coordinate system origin is at the D2 mechanical center and the units are in centimeters. The mechanical centers of D3 and D4 are at (-0.772, 0, 181.4) and (-9.155, 0, 384.3); the center of the IDSM entrance face is at (19.40, 2.3, 493.1), about which the magnet is rotated by 2.613°. The coordinates of the two stripping foil centers are (3.997, 2.3, 30.71) and (7.29, 2.3, 293.5). All the mechanical dimensions are taken from the BNL design parameters and are verified by survey and alignment data [10]. Note that the two foil centers are 2.3 cm above the mid-planes of the three chicane dipoles. The D4 axis is offset from the D2 axis by -9.155 cm in the x-direction. These component locations have significant consequences for particle motion.

We studied all the three sets of chicane dipole settings in different three-dimensional models and compared their performance. The preliminary investigation of the waste beam losses in a model with the delivered setting was reported in [11]. Here we present the results only from the production setting. This setting was employed during the ring commissioning and early operations, and typically showed all the waste beam loss mechanisms [12, 13]. In the production setting the current amplitudes of chicane dipoles D2 and D3 are 2126

A and 1449 A, while they are 1737 A and 2914 A in D4 and IDSM, respectively. These currents in D2, D3, and D4 yield bending angles of -53.1, -28.3, and 39.4 mrad for a 1 GeV proton beam. Chicane dipole D1, which is not a part of the waste beam line and is not included in the models, has a bending angle of 42.0 mrad in all the three chicane settings.

A great difficulty with these three-dimensional simulation models is their very large volumes, which present a significant challenge in mesh generation. All the cell properties must be carefully assigned and adjusted. A number of artificial cuts must be inserted in the models in order to make fine meshes. A model typically contains about 14 million total elements, which are close to the maximum allowed by OPERA-3D version 11 [14]. Other model statistics include 8.3 million total nodes, 16 million edges, 9 million linear tetrahedral, 5 million quadratic tetrahedral, and 8 million equations. The TOSCA solutions of the models yield a post-processor file of 4.132 GB for each setting, from which we obtain the field distributions and three-dimensional particle trajectories.

8.3.2. Magnets and Fields on Beam Line

All the magnets for the SNS ring injection system and its waste beam line were designed and developed by the BNL/SNS team. A great effort was devoted to the modeling, design optimization, and measurements of these devices at BNL. We have also performed three-dimensional simulations of these magnets individually with OPERA-3D/TOSCA, and obtained quite good agreement with the results reported by BNL. The main design features of these magnets and field distributions are described below. This helps to understand the waste beam line operation.

Chicane dipoles D2 and D3 have been studied extensively in three-dimensional simulations and experimental measurements [5, 7, 8, 15]. These dipoles are C-shaped magnets with complementary pole tips. The two dipoles have no symmetric bending planes, thus the magnetic field is three dimensional everywhere. The primary foil (F1) is located around the D2 downstream edge, where the dipole field B_y is 2.5 kG. The field tilt (Atan $(-B_z/B_y)$) at F1 is about 220 mrad. The ratio of the D2 field integrals upstream and downstream of the foil is $R = 3.784$. The peak field of D3 is less than 1.9 kG. The secondary foil (F2) is located downstream of dipole D3. Figure 8.3 shows the field distribution along a straight line from the F1 center (3.997, 2.3, 30.71) to the F2 center (7.29, 2.3, 293.5). The B_y integral is 0.2201 T m, yielding a total bending angle of 38.9 mrad for a 1 GeV H^- particle.

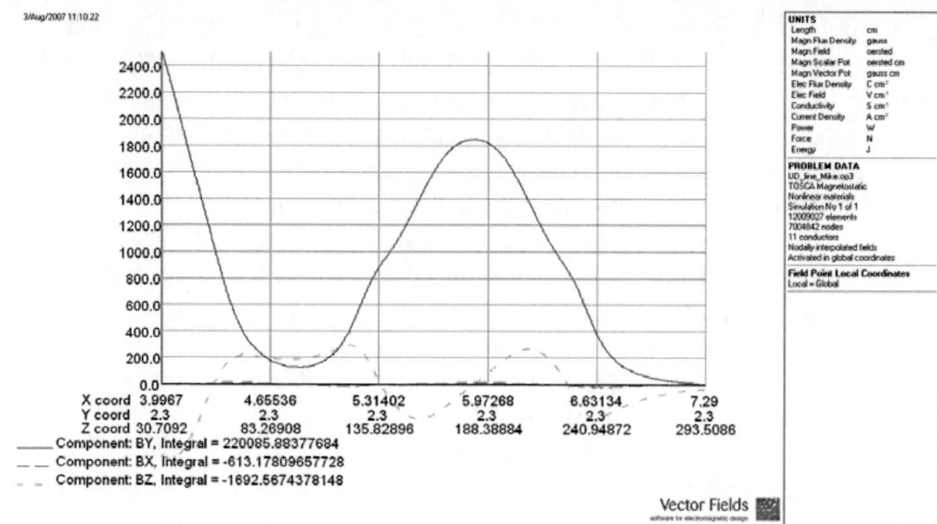

Figure 8.3. Magnetic field along a line from F1 center to F2 center, where the units are centimeter for x, y, z coordinates and gauss for magnetic flux density.

Chicane dipole D4 is a typical H-shaped magnet [5, 16]. The dipole bends both the circulating beam and the waste beams, so it should have a large horizontal aperture with good field quality. The pole tip is 51.82 cm wide. In order to improve the field quality, the pole-tips have longitudinal shims on their edges, and the magnet also contains eight z-bumps to enhance its field in the region far from its center. Although the dipole has mid-plane symmetry, most waste beam particles enter the magnet already above the symmetry plane. Thus, the fields for all the reference trajectories are three dimensional.

The injection dump septum magnet [5, 16, 17] is a sector dipole with significant quadrupole components. Its curved pole-tips are 193 cm in length and 26.52 cm in width. The septum handles both the H^0-proton and H^--proton waste beams with their trajectories going through two sides of its longitudinal axis. The vertical aperture of its vacuum chamber is as small as ± 2.07 cm. Although the magnet has up-down symmetry, most waste beam particles coming from D4 are already above the symmetry plane. Figure 8.4 shows the magnetic field distribution on the mid-plane across the IDSM gap at its longitudinal center. The two vertical, dashed lines indicate the pole-tip width boundaries, while the two solid lines depict the relative positions of the vacuum chamber inner surface. The B_y plot indicates that the dipole has a gradient, which provides horizontal focusing and vertical defocusing for the waste beams. The average slope of the B_y curve on the left side of the pole-tip center is about 0.97 T/m, while it is 1.14 T/m on the other side. Thus, the H^--proton beam is focused stronger in the x-direction than the H^0-proton beam.

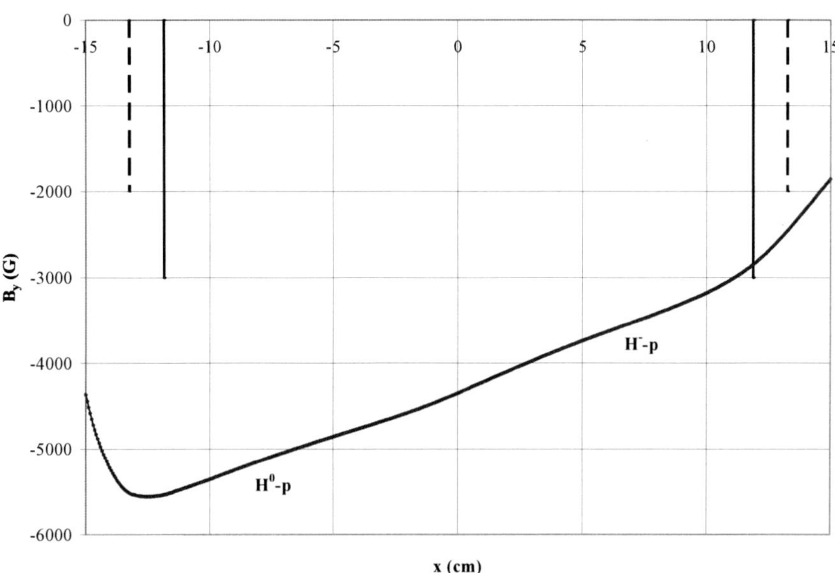

Figure 8.4. Magnetic field on the mid-plane across the IDSM gap at its longitudinal center.

The magnetic field component B_y along two selected waste beam trajectories on the mid-plane is illustrated in Figure 8.5, where z = 0 indicates the IDSM entrance face. Both the H^0-proton and H^--proton tracks are obtained by launching 1 GeV particles upstream in perpendicular to the IDSM entrance face. The H^0-proton starts at 9 cm from the IDSM axis, while the H^--proton starts at 4 cm on the other side of the axis. We pick these positions since the chicane dipole settings usually make the two waste beams asymmetric with respect to the IDSM axis. The total bending field integral along the H^0-proton track is 1.034 T m corresponding to a bending angle of 182.8 mrad, while it is 0.7898 T m, or 139.6 mrad, for the H^--proton track. These bending angles are fairly close to the values required by conditions (8.5a) and (8.5b). In reality, both the waste beams, especially the H^--proton particles, have large incident angles with respect to the IDSM entrance face. Thus, the actual trajectories would deviate appreciably from these selected tracks, and the magnetic fields along real beam trajectories could be quite different from what is shown in the figure.

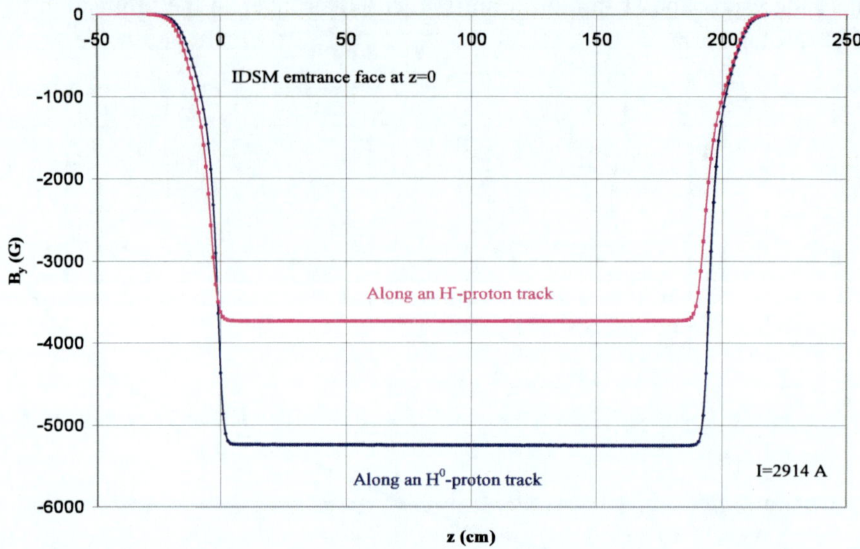

Figure 8.5. Magnetic field along two selected waste beam trajectories on IDSM mid-plane.

8.3.3. Initial Conditions of Test Particles

We take a test particle approach in our calculation of the waste beam particle trajectories. The primary stripping foil F1 is nominally 12 mm wide and is mounted at an angle of 30° with respect to the incoming H^- beam. Thus, the incoming beam only sees a foil width of ± 5.2 mm as shown in Figure 8.6(a), which is the view towards upstream. Measurements show that the incoming H^- beam has a slightly elliptical profile with its long axis of about 15 mm in the y-direction. We use a round beam profile for the convenience of calculation, and thus we slightly overestimate the waste beam dimension in the x-direction. The H^0 particles are generated in the foil area, and we pick three of them each on the left (blue diamonds) and right (red diamonds) edges and one each on top (purple diamond) and bottom (green diamond). The majority of the H^- particles after F1 should come from outside the foil, i.e., the three island areas. We pick four of them each on the left (blue diamonds), right (red diamonds), and bottom (green diamonds). Particles with their initial velocity parallel to the z-axis are represented by solid diamonds in our plots later. The particles from the F1 center (3.997, 2.3, 30.71) are shown by a magenta dot for the H^0 centroid and a magenta circle for the H^- centroid.

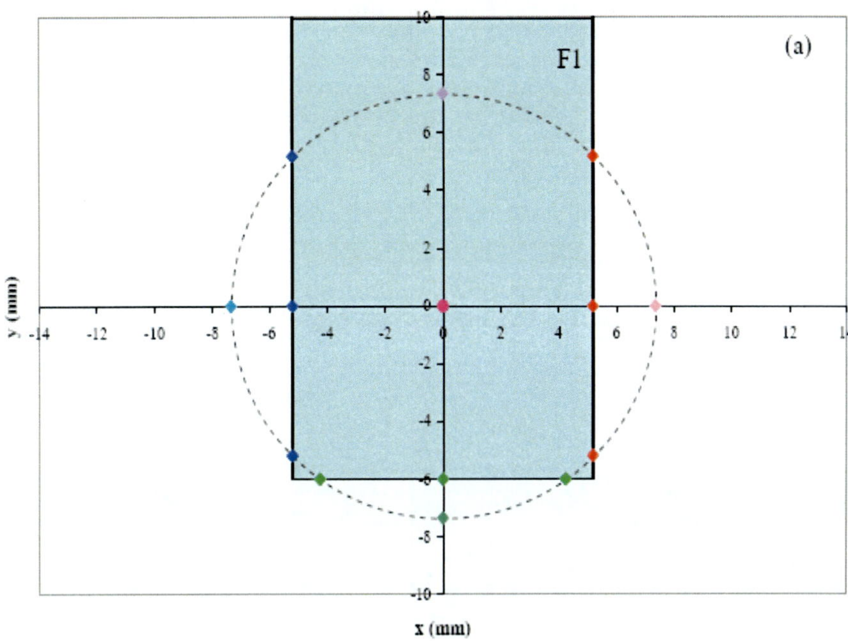

Figure 8.6. (a) Initial test particles in configuration space.

The assignment of initial transverse angles for the test particles is based on the following considerations. The normalized rms emittance of the injected H^- beam is 0.5 π mm mrad and the rms radius is about 1.5 mm [9]. Since the beam has halo tails, its profile is not a true Gaussian, and it occupies much larger area in the real space than expected from a Gaussian distribution. Thus, we employ the concept of equivalent beams [18], which satisfy the Boltzmann-Maxwell distribution in phase space. This beam has an effective rms radius of 2.1 mm and an un-normalized rms emittance of 0.28 π mm mrad. Figure 8.6(b) illustrates two different phase space boundaries for this beam: one-rms (plum dashed ellipse) and twelve-times rms emittance (black dashed ellipse). A beam with the Boltzmann-Maxwell distribution in phase space has only 0.25% particles outside the twelve-times rms emittance area. This provides a reasonably good criterion for cut-off in the calculations. The particles launched at the twelve-times rms emittance boundaries with non-zero initial angles are denoted by crosses as shown later in Figures 8.10, 8.11, etc. The scattering in F1 is negligibly small and is ignored. The scattering in F2 may play some role in spreading particle distributions and causing particle losses, but that effect is not included in these calculations.

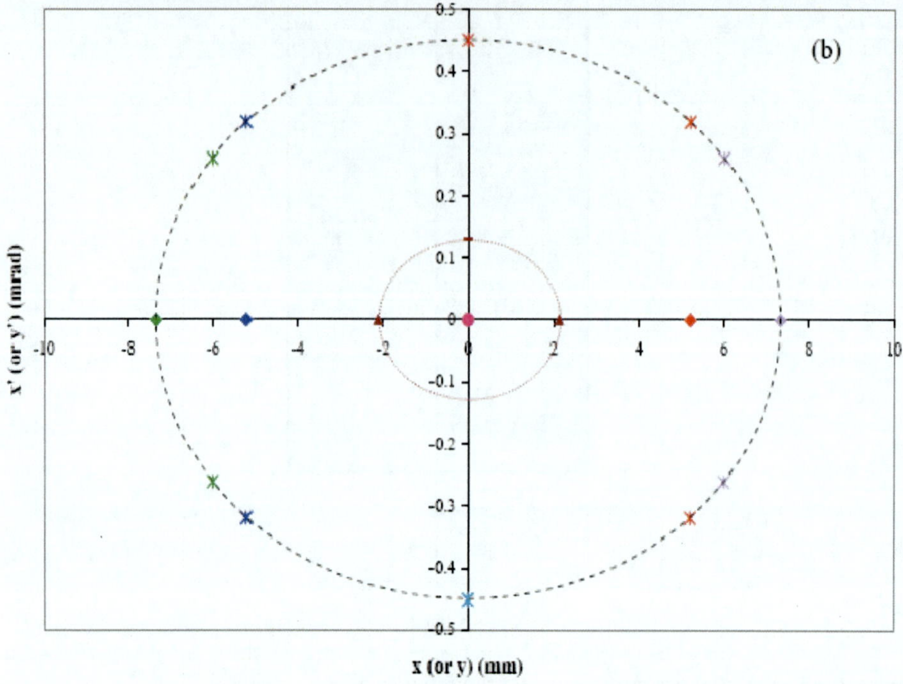

Figure 8.6. (b) Initial test particles in phase space.

8.4. Three-Dimensional Particle Trajectories Through IDSM

OPERA-3D has a built-in TRACK command for calculating trajectories of charged particles by directly integrating the equations of motion in simulated magnetic fields [19]. We launch waste beam particles of 1 GeV in energy at F1 with different initial locations and slopes as specified in Figures 8.6(a) and 8.6(b). The H^0 particles do not feel any force in between F1 and F2. They simply drift in the region and the calculation of their trajectories actually starts at F2. For the H^- particles, we first compute their trajectories in between F1 and F2, and obtain their positions and velocities at F2. This information is used as the initial conditions of the waste H^--proton beam at F2. The new Euler's angles at F2 are phi = Atan(v_y/v_x), theta = Atan($v_\perp/v_z$), and psi = 0, where $v_\perp$ is the transverse velocity of the H^- particles arriving at F2. These angles are required by the OPERA-3D TRACK command.

Figure 8.7 shows the H^- particle trajectories (in between F1 and F2) and H^--proton particle trajectories (after F2) in the y-z plane. Three chicane dipoles D2, D3, and D4 in the production setting are depicted by their axial positions, and the IDSM is indicated by the inner surface of its vacuum chamber. The simulation model extends only to z = 1000 cm, beyond which the particle trajectories are obtained by linear extrapolation. The trajectory plot stops at a z-position marked by a light blue cross, which is 1.3 m from the downstream quadrupole magnet center. It is clear that all the H^--proton particles have significant y-motion in D4 and IDSM, and some H^--proton particles are lost to the upper surface of the IDSM vacuum chamber. These losses are much more serious in the design setting and the delivered setting of the chicane dipoles.

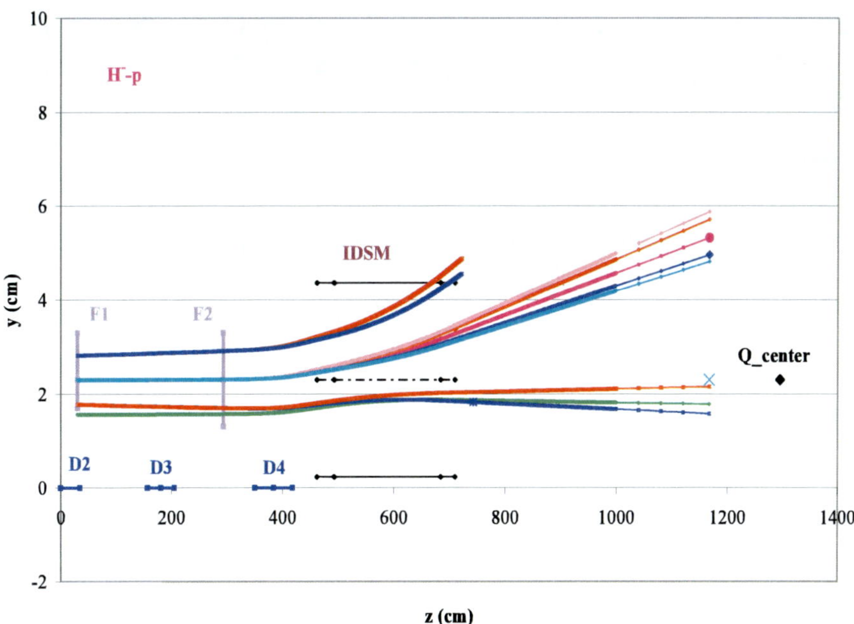

Figure 8.7. H^- and H^--proton particle trajectories in the y-z plane.

Figure 8.8 shows the H^0 particle trajectories (in between F1 and F2) and H^0-proton particle trajectories (after F2) in the y-z plane. In comparison with Figure 8.7, the y-motion of all the H^0-proton particles in the D4 region is unnoticeable, and is still quite moderate in the second half of the IDSM. We do not see any losses of these particles in the y-direction. This is also true for the other two settings.

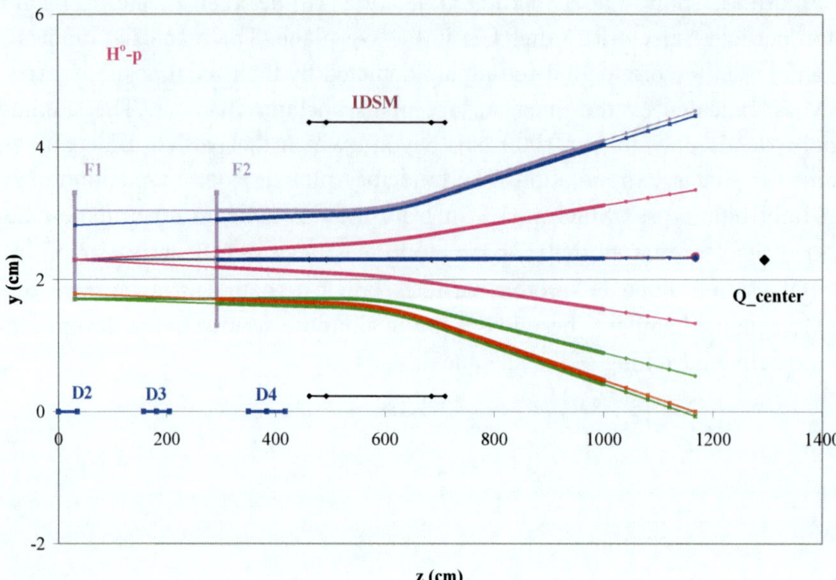

Figure 8.8. H^0 and H^0-proton particle trajectories in the y-z plane.

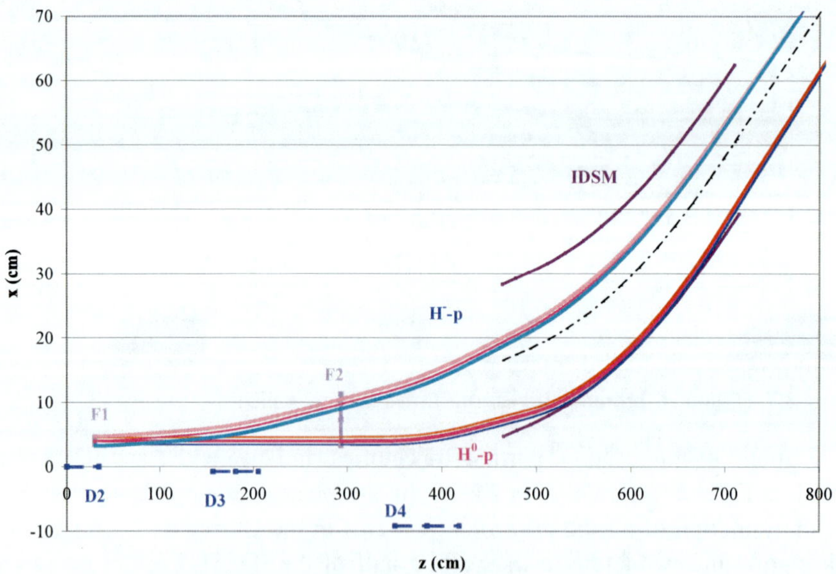

Figure 8.9. Waste beam trajectories in the x-z plane.

Figure 8.9 shows the waste beam trajectories in the x-z plane. The two bundles of the H^--proton and H^0-proton beams include their centroids and two more tracks in the maximum x-extension. We can see that the two waste beams are very asymmetric with respect to the IDSM axis. The H^0-proton beam is too close to the IDSM vacuum chamber wall and some particles already hit the middle of the IDSM vacuum chamber. This is consistent with what was observed in early operations. These horizontal losses for the H^0-proton particles are not seen in the two other settings since they have stronger bending angles in D4. This will be explained in more detail in Section 8.6.2.

The waste beam trajectories in the IDSM shown in Figure 8.9 are quite different from the selected tracks used for the B_y plot in Figure 8.5, since their initial conditions before the IDSM entrance are quite different. Calculations show that in the IDSM coordinates, where the origin is at the center of the IDSM pole-tip entrance face, the H^0-proton centroid in Figure 8.9 arrives in front of the septum at x = -10.9, y = 0.0033, and z = -30.7 cm with the incident angles of $\mathrm{Atan}(v_x/v_z)$ = -6.5 mrad and $\mathrm{Atan}(v_y/v_z)$ = 0.022 mrad. The H^0-proton centroid is too far away from the IDSM longitudinal axis; and the negative incident angle of -6.5 mrad in the horizontal direction drives it further toward the vacuum chamber wall. Note that the vacuum chamber has a horizontal aperture of only ± 11.85 cm. Thus, it is easy to understand that some H^0-proton particles around their centroid will hit the vacuum chamber wall somewhere inside the IDSM.

On the other hand, the H^--proton centroid arrives in front of the IDSM at x = 1.29, y = 0.186, and z = -30.4 cm with the incident angles of $\mathrm{Atan}(v_x/v_z)$ = 30.1 mrad and $\mathrm{Atan}(v_y/v_z)$ = 2.3 mrad. Although the H^--proton centroid is initially very close to the IDSM pole-tip center, its large incident angle of 30.1 mrad causes it to gradually deviate more and more from the IDSM axis. This mismatch between the IDSM design parameters and initial conditions makes it very difficult to transport the two waste beams through the IDSM, as well as through the rest of the way to the dump.

Figure 8.10 shows all the waste beam particles at the IDSM entrance looking upstream. The plot is made by the interceptions of the two waste beam particles on the IDSM input surface. The blue dashed line indicates the IDSM mid-plane position. The pattern of the H^0-proton test particles is very similar to that at F1. However, the H^--proton particles already gain significant y-motion, and their centroid is Δy = 0.26 cm above the IDSM mid-plane. The horizontal distance between the H^0-proton centroid and the H^--proton centroid is about 13 cm. The focusing of the H^--proton particles in the x-direction and defocusing in the y-direction can also be seen. This is due to the edge focusing in the D4 fringe field.

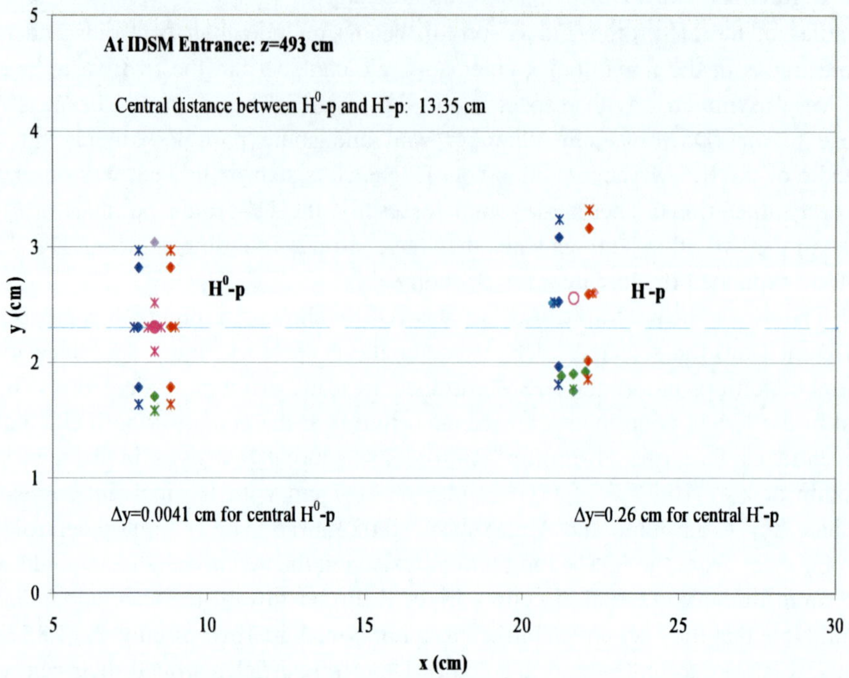

Figure 8.10. Waste beam particles at the IDSM entrance.

Figure 8.11 shows all the waste beam particles at the IDSM chamber exit, again looking upstream. The H^--proton particles centroid is $\Delta y = 0.98$ cm above the mid-plane and some of them are already outside the vacuum chamber. The horizontal distance between the H^0-proton centroid and the H^--proton centroid is increased to about 15 cm. However, the H^0-proton particles are very close to the chamber wall in the horizontal direction. This is consistent with the plot in Figure 8.9. The vertical separation and horizontal distance between the two waste beam centroids are important parameters in beam measurements as explained in Section 8.7. In Figure 8.11, we can also see that the H^--proton particles are much more strongly focused in the x-direction and defocused in the y-direction than the H^0-proton particles. This is not only due to the higher gradient term inside the IDSM for the H^--proton particles (Figure 8.4), but also due to much stronger edge focusing/defocusing of the H^--proton particles, which are incident to the IDSM with large face-rotation angles.

In the delivered setting and the design setting of the chicane dipoles, there are even more H^--proton particles outside the vacuum chamber at its exit position as in Figure 8.11. More H^--proton particles are lost on the upper surface of the septum vacuum chamber. On the other hand, the H^0-proton particles in these two settings have good distances from the vacuum chamber wall in the horizontal direction, and for these cases the H^0-proton particle losses in the septum may not be a problem.

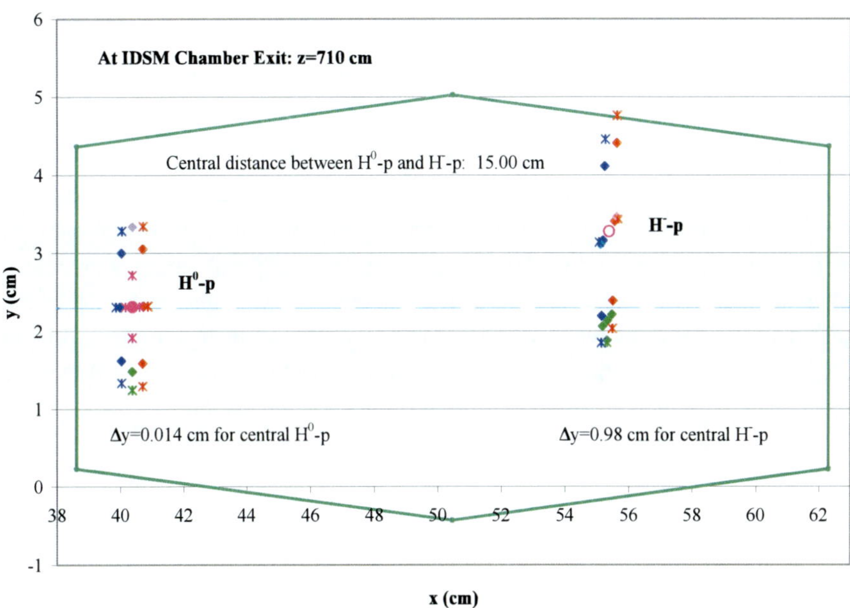

Figure 8.11. Waste beam particles at the IDSM chamber exit.

8.5. Particle Optics Through Quadrupole Magnet to Dump

The waste beam transport after IDSM through the quadrupole magnet to the injection dump is shown in Figure 8.12. The region from z = -1.3 to z = 1.3 m is for the quadrupole magnet 30Q58, including its fringe field, where the vacuum chamber aperture radius is 9.86 cm. The remaining is just a drift space with the shielding wall at z = 9.087 m and the injection dump window at 19.5 m.

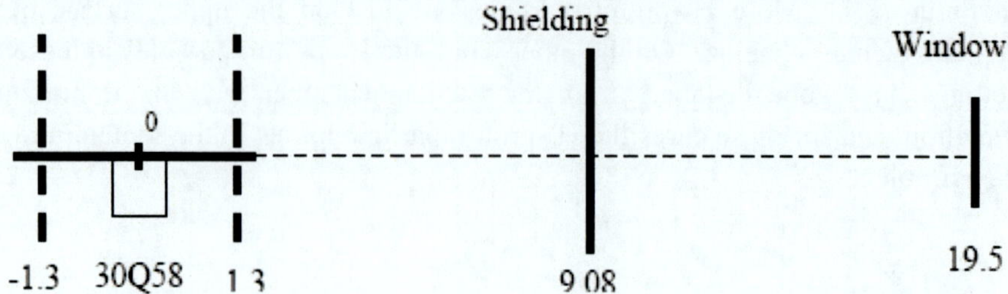

Figure 8.12. Waste beam transport through a quadrupole magnet to dump (dimension in meters).

For the particle optics analyses, we first transform the information in the previous section to the quadrupole magnet frame to obtain the initial conditions at z = -1.3 m. Figures 8.13(a) and 8.13(b) show the waste beam phase space distributions in the x-x′ and y-y′ phase space at the quadrupole magnet entrance. It is clear that the two waste beam particles occupy very large areas in the phase space. It is also easy to see that the waste beam is converging in the x-direction and diverging in the y-direction, requiring a horizontally defocusing quadrupole magnet.

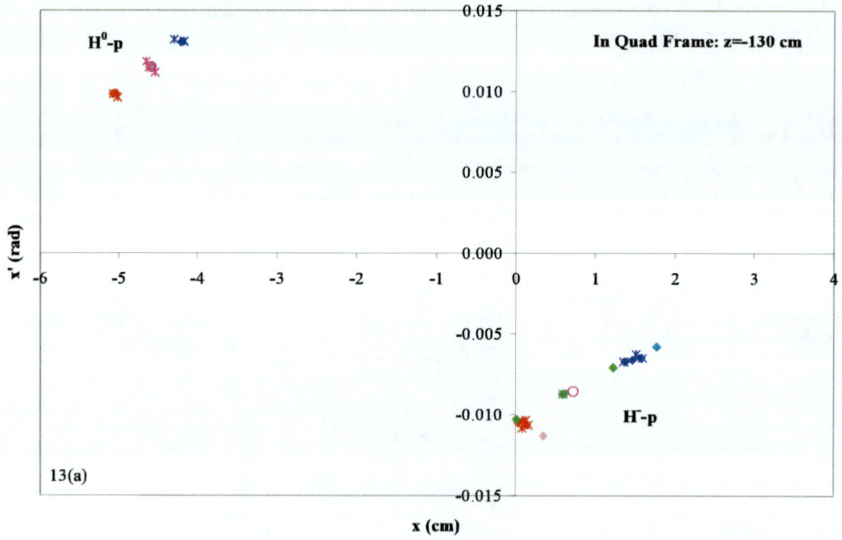

Figure 8.13. (a) Phase space distribution of waste beams in x-x′ space at quadrupole magnet entrance.

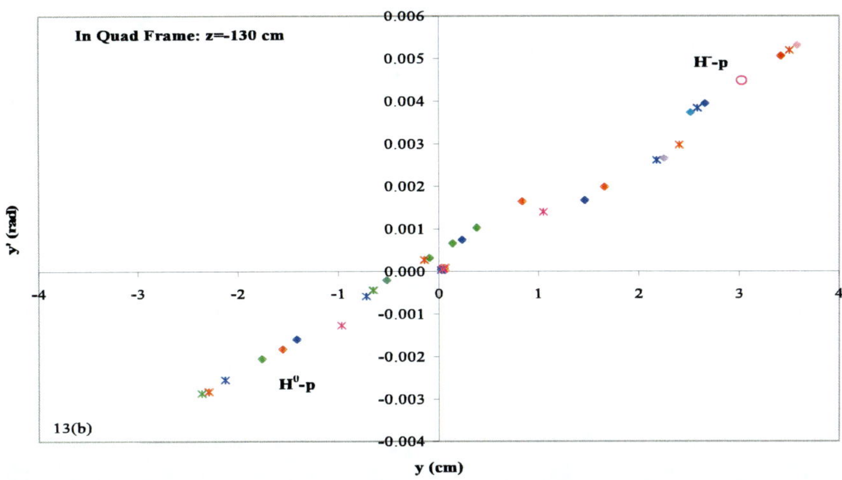

Figure 8.13. (b) Phase space distribution of waste beams in y-y′ space at quadrupole magnet entrance.

The quadrupole magnet 30Q58 was originally designed for the ring straight sections, and its performance characteristics were well studied [20, 21]. Instead of a simple hard edge model for the quadrupole magnet, we employ its transfer matrices rigorously computed from the equations of motion with the fringe field effect, as described in Chapter 5. We first use the transfer matrices to map the waste beam particles from z = -1.3 to z = 1.3 m, and then to use the drift space matrices to map the particles further downstream. A number of different quadrupole magnet currents are used in calculations and the best result is shown in Figure 8.14, where the quadrupole magnet parameters are listed in Table 8.2, and the green circle indicates the inner surface of the vacuum chamber. A single quadrupole magnet 30Q58 can not even transport all the waste beam particles through the shielding wall, let alone through the dump window which is more than ten meters downstream. The situations are even worse for the design setting and the delivered setting.

Table 8.2. 30Q58 parameters at I = 405 A.

	m_{11}	m_{12} (m)	m_{21} (1/m)	m_{22}	f (m)
F	0.7240	2.2508	-0.2114	0.7240	4.731
D	1.2893	2.9638	0.2235	1.2893	-4.475

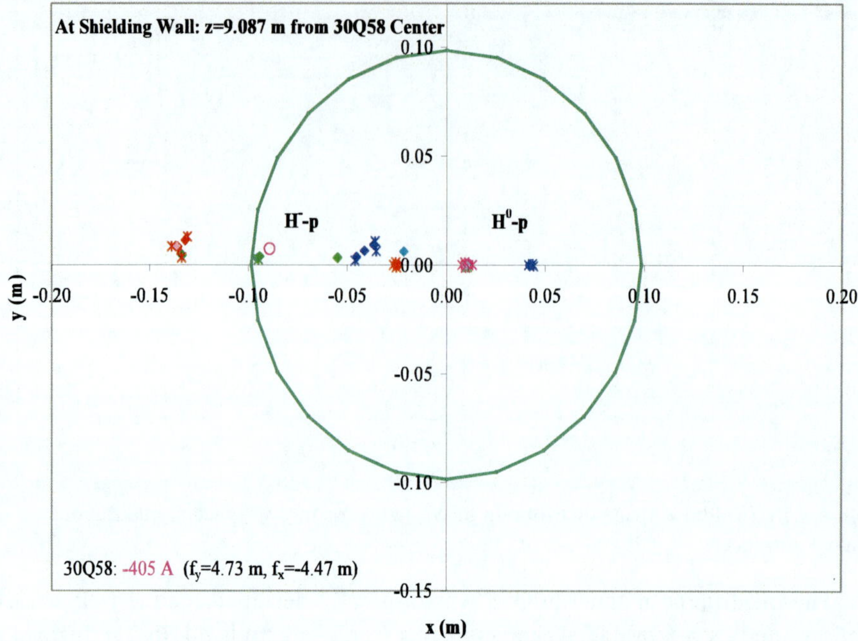

Figure 8.14. Waste beam particle positions at shielding wall for a quadrupole current of -405 A.

The injection dump optics in the original design [8, 22] shows that the beta function for both x and y at the dump window reaches approximately 3500 m. With a 10″ pipe for the transport channel, the acceptance is about 4.4 π mm mrad. Although this acceptance is significantly larger than the rms emittance of the linac injected H^- beam, it is far less than the equivalent phase space area occupied by both H^0-proton and H^--proton waste beam particles, as shown in Figures 8.13(a) and 8.13(b). This makes the waste beam transport downstream to the dump window impossible.

8.6. REMEDIES

The studies of three-dimensional particle trajectories in the SNS ring injection dump beam line have shown three waste beam loss mechanisms in the IDSM and downstream. First, the H^--proton particle motion in the y-direction is excessive and the vertical aperture of the IDSM vacuum chamber is not adequate. Many H^--

proton particles are lost on the upper surface of the IDSM vacuum chamber. This is true for all three sets of chicane dipole settings. Second, many H^0-proton particles are lost in the horizontal direction in the middle of the IDSM for the production setting. This is mainly caused by insufficient bending field strength in chicane dipole D4 in the production setting. A larger horizontal aperture of the IDSM will also reduce the H^0-proton particle losses. For the other two settings, i.e. the design and delivered settings in Table 8.1, these losses do not appear in the simulations since they employ stronger bending field in D4. Third, by employing only a single quadrupole magnet, the waste beams can not be fully transported through the shielding wall in any of the three chicane dipole settings.

After more detailed investigation of these beam loss mechanisms through three-dimensional modeling, we have devised solutions to the problems as described below.

8.6.1. H^--Proton Particle Losses in the Y-Direction in IDSM

The vertical motion of the H^--proton particles in the chicane region is mainly caused by chicane dipole D4 as shown in Figure 8.15, which is calculated for the production setting. This is because the H^--proton trajectories pass through the D4 pole-tip boundary, where the magnetic fields are very non-uniform and there is a significant B_x component. Figure 8.16 shows the geometry in the D4 region, where the D4 axis is at x = -9.155 cm and another half of dipole D4 is not included in the figure. The magnetic fields along the H^--proton centroid trajectory through D4 are plotted in Figure 8.17. The integrated B_x is about 6% of the B_y integral for the production setting. The ratio of the B_x integral over the B_y integral through D4 is more than 10% for the delivered setting. It is this strong B_x integral that pushes the H^--protons in the y-direction. In contrast, the circulating proton beam and the H^0-proton waste beam go through the D4 central region, where the fields are quite uniform and the component B_x is very small. Thus, their vertical motion is very moderate there.

If we slide D4 in the positive x-direction, the H^--proton tracks would move toward the D4 center into a more uniform field region. The B_x component is then much reduced, and the y-motion of the H^--proton particles could be greatly mitigated. The minimum movement is $\Delta x = 8$ cm due to mechanical constraints, though a D4 movement of $\Delta x = 4 \sim 6$ cm would suffice according to simulation studies.

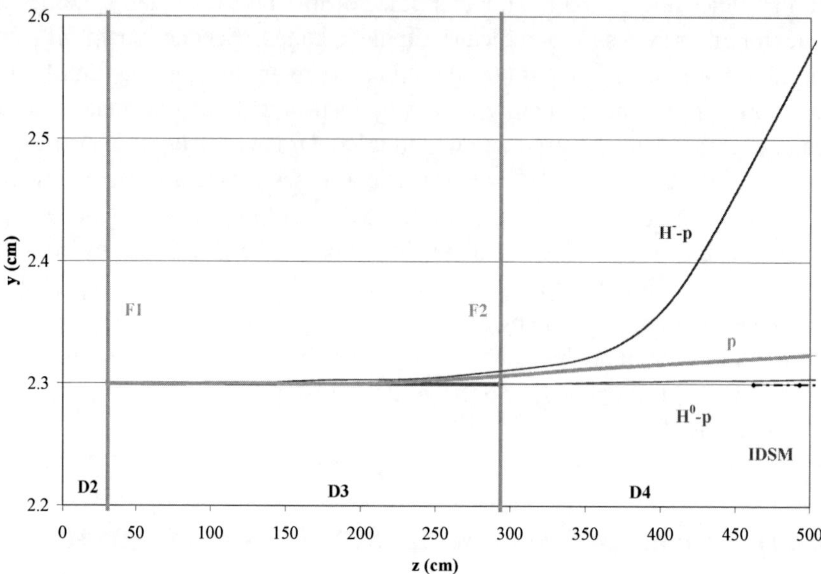

Figure 8.15. Waste beam particle trajectories in the y-z plane in the region of chicane dipoles.

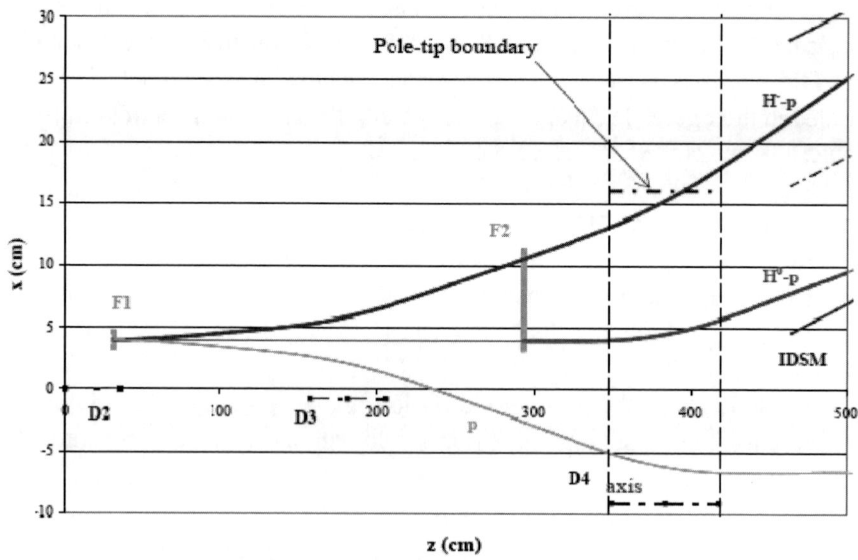

Figure 8.16. Particle tracks in the D4 region and vicinity in the x-z plane.

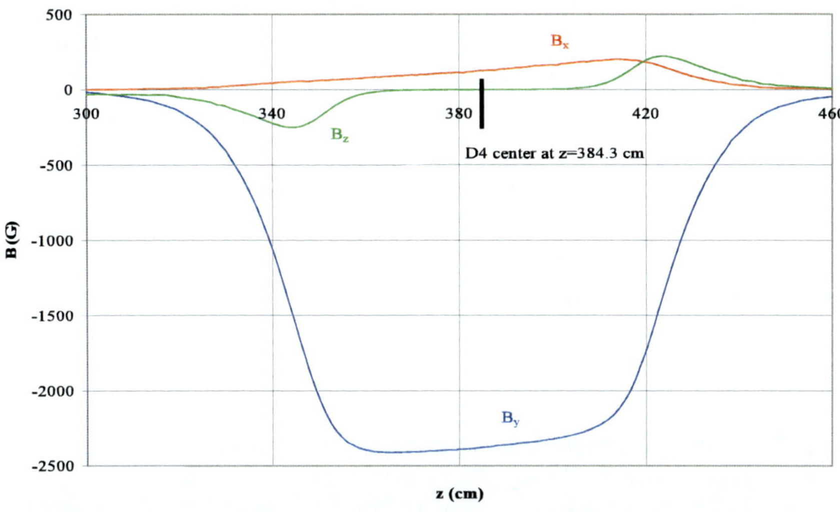

Figure 8.17. Field distribution along the H⁻-proton centroid track in the D4 region.

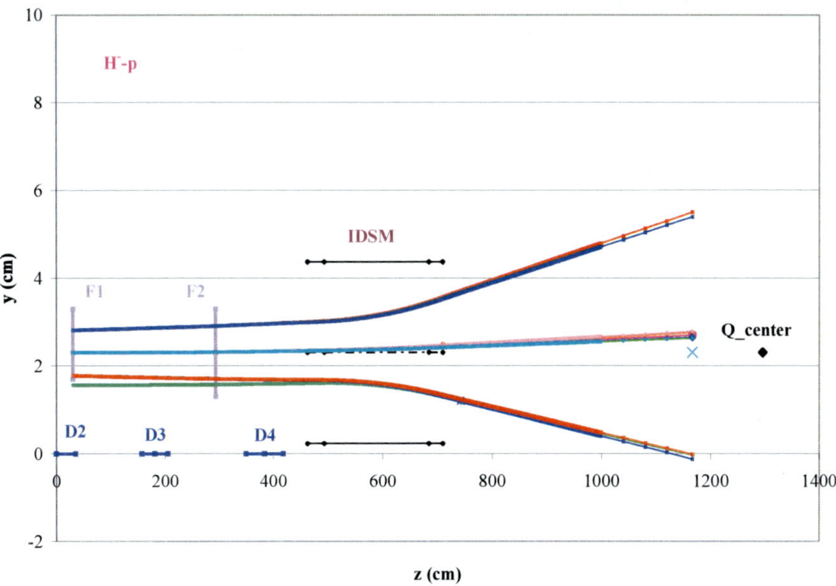

Figure 8.18. H⁻-proton particle trajectories in the y-z plane after the D4 move by $\Delta x = 8$ cm.

We have built new simulation models in which the D4 center is moved from the original x = -9.155 to x = -1.155 cm (Δx = 8 cm). Figure 8.18 shows the y-motion of the H^--proton particles in the new configuration under the production setting. All the particles remain inside the IDSM, in contrast to that in Figure 8.7. This has also been verified for the other two settings. Thus, moving D4 by Δx = 8 cm would significantly reduce the y-motion of the H^--proton particles as well as their losses in the IDSM. On the other hand, the move would push the circulating proton beam towards the non-uniform field region in D4. This has been checked in a separate simulation, which shows unnoticeable effect on the circulating proton beam after D4 is moved by Δx = 8 cm [23].

8.6.2. H^0-Proton Particle Losses in the X-Direction in IDSM

The waste beam trajectories in the IDSM are very sensitive to the chicane dipole settings. With the production setting where the D4 field strength is smaller than that in the other two settings, many H^0-proton particles would be lost in the horizontal direction around the middle of the IDSM. Since the IDSM can not be moved in the minus x-direction, a possible remedy for this problem is to move the injection point at F1 in the positive x-direction. This leads to the H^0-proton tracks off the IDSM vacuum chamber wall.

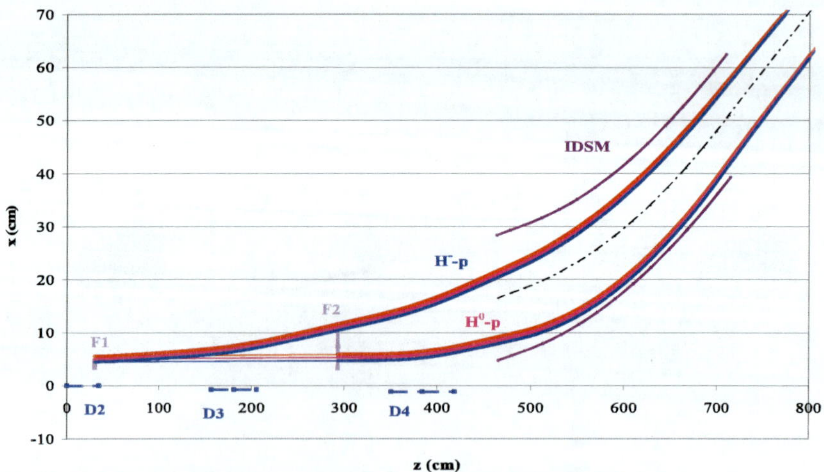

Figure 8.19. Waste beam trajectories in the x-z plane after F1 move by Δx = 1 cm and with injection at 1 mrad.

Figure 8.19 shows the waste beam particle trajectories in the x-z plane for the production setting model with F1 moved by $\Delta x = 1$ cm and injection at 1 mrad angle with respect to the z-axis. It appears that this is a reasonable solution. This also requires an adjustment of the dynamic bump amplitude by +1 cm to keep the painting parameters unchanged. Moreover, the F1 longitudinal position should be slightly ahead of the static bump peak in order to accommodate the 1 mrad injection angle for good injection.

An alternative way to accomplish the same task is to develop a new chicane dipole setting, in which a stronger bending field in dipole D4 should be provided in addition to meeting other injection constraints. To accommodate the F1 move by $\Delta x = 1$ cm, we can increase the static bump height to 10.66 cm from the original 9.66 cm as calculated in Section 8.2.1. The primary foil F1 will stay at the peak of the static bump. In order to inject at right position with zero angle with respect to the z-axis, the incoming H^- beam line has to be adjusted as well in parallel with the z-axis at F1. This requires fine tuning of the injection line. From Eqs. (8.1) – (8.4) we obtain the following bending angles: 46.35, -58.60, -31.19, and 43.44 mrad for chicane dipoles D1, D2, D3, and D4, respectively. This setting provides good injection and guarantees a closed injection bump and a good equilibrium orbit for the circulating proton beam. In addition, the new D4 bending angle (43.44 mrad) is larger than that in the production setting (39.4 mrad). This is good for the H^0-proton beam in terms of its clearance from the IDSM chamber wall in the horizontal direction.

We have built a three-dimensional simulation model under the new chicane dipole setting with the D4 position already moved by $\Delta x = 8$ cm. The calculations have been done with the initial centroid particle at $x = 4.997$ cm, i.e. $\Delta x = 1$ cm from that in the original production setting. Figure 8.20 shows the waste beam particle trajectories in the x-z plane. The H^0-proton particles are kept clear of the IDSM vacuum chamber wall. In comparison with Figure 8.9 for the production setting, this is a much better transport for the waste beam particles through the IDSM. In the new chicane dipole setting with D4 moved by $\Delta x = 8$ cm, both the H^0-proton beam and the H^--proton beam go through the IDSM without showing any losses in the y-direction.

Figure 8.21 shows all the waste beam particles at the IDSM chamber exit, again looking upstream. The H^--proton centroid is $\Delta y = 0.26$ cm above the mid-plane, as indicated by the dashed blue line, and it is about 16.9 cm from the H^0-proton centroid. The two bunches of the waste beam particles are symmetrically distributed in the horizontal direction with respect to the IDSM center. All the waste beam particles stay within the chamber, though the clearance is still tight.

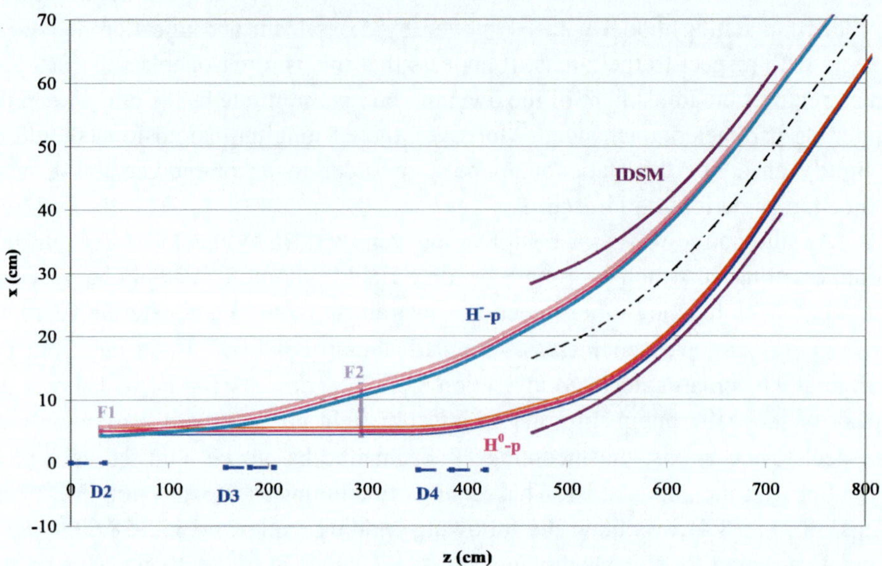

Figure 8.20. Waste beam trajectories in the x-z plane under the new chicane dipole setting.

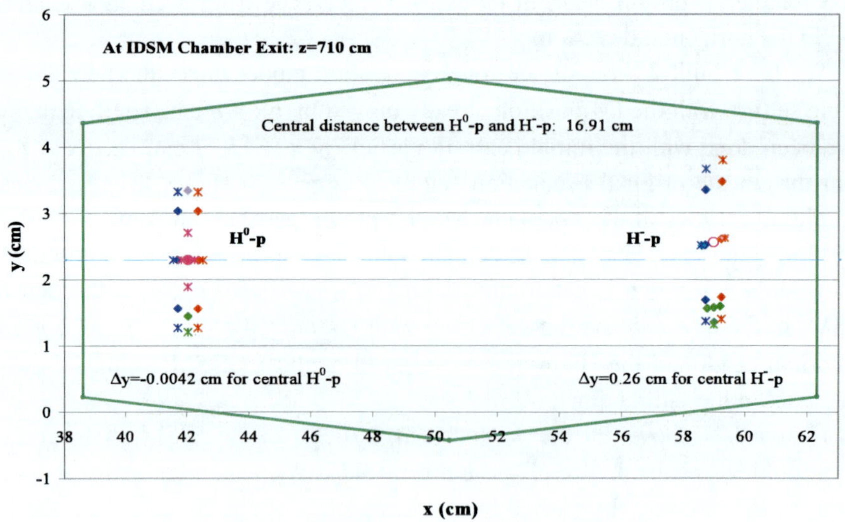

Figure 8.21. All waste beam particles at the IDSM chamber exit in the new chicane dipole setting.

8.6.3. Modification of a Spare IDSM

The IDSM vacuum chamber aperture is ± 11.85 cm in the horizontal dimension and is as small as ± 2.07 cm in the vertical dimension. This is inadequate to transport the waste beams even after we make the changes described in the previous two subsections. We have modified a spare IDSM for the improvement of its performance [24]. First, the vertical gap of the magnet is increased by 2 cm. This provides both the H^--proton and H^0-proton particles more clearance in the y-direction. Second, good field region for the H^0-protons is further extended from the pole-tip edge toward the septum plate, where it happens that there is some space available. Thus, the vacuum chamber can be accordingly enlarged horizontally by almost 3 cm, which gives the H^0-proton particles significantly more clearance in the x-direction. Third, we add specially designed z-bumps on the IDSM entrance face for the H^0-proton beam that further pushes these particles inward in the IDSM, and reduces the distance between the H^0-proton centroid and the H^--proton centroid at the IDSM exit.

Figure 8.22 shows the IDSM vacuum chamber inner surface before and after the modification. The new chamber has a trapezoidal cross section that yields more vertical space on one side. Both the waste beams gain more clearance in the vertical direction with the H^--protons benefited more since they pass through the right side of the chamber. The H^0-proton beam gains exclusively more clearance in the horizontal direction that reduces their losses inside the IDSM.

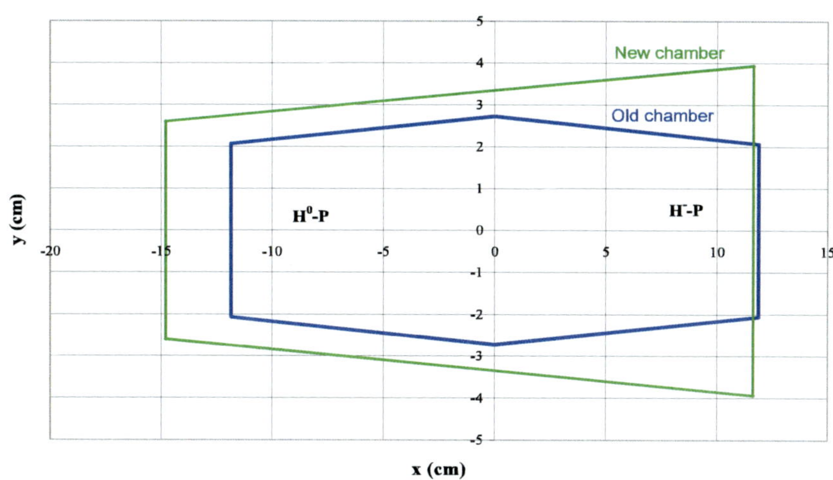

Figure 8.22. IDSM vacuum chamber inner surface dimensions (upstream view).

Figure 8.23 shows the magnetic bending field B_y across the magnet gap on the mid-plane at the center of the magnet before and after the modification. The longitudinal magnet axis goes through x = 0. The two black vertical lines indicate the positions of the old vacuum chamber inner surface. The two black dashed lines are the pole tip boundaries of the old magnet. The new vacuum chamber inner surface on the left side (red vertical line) is 2.95 cm away from the old one, and the new pole tip boundary (red dashed line) is 1.2 cm away from the old one. The dark blue curve is the same as that in Figure 8.4 for the old septum which is operated at 2914 A. The red curve is from the modified magnet model, which would operate at a current of 3800 A. Although the field strength on the left side (for the H^0-protons) is smaller in the modified septum, the integrated field is about the same as the old one since the new z-bumps are added on this side. The main disadvantage of the modification is that the modified septum has rather low gradient for the H^0-proton particles due to the enlarged gap.

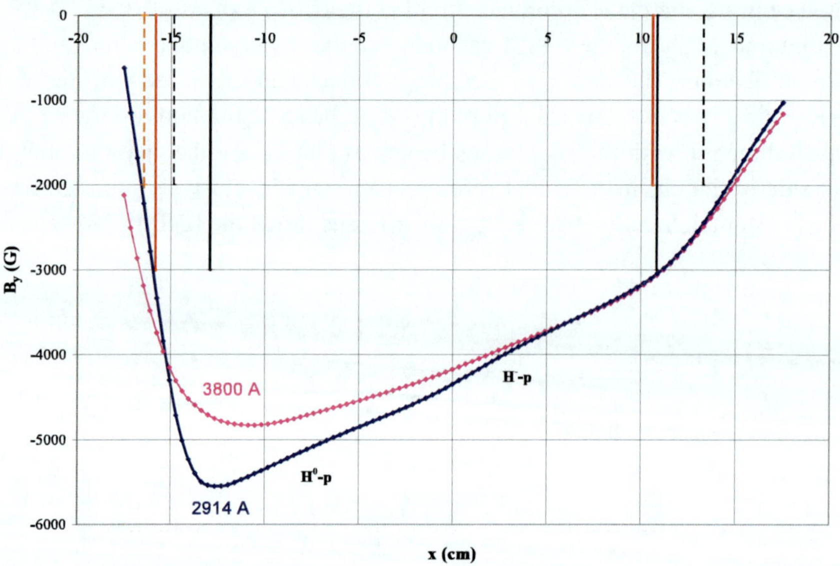

Figure 8.23. Magnetic field B_y across the gap on the mid-plane at the septum center.

Figure 8.24 shows the two waste beam centroids in the x-z plane. The initial conditions in the front of the septum are determined from the production chicane setting after D4 is moved. We compare two sets of particle trajectories: one is for the modified septum and the other is from the old one. The H^--proton centroid trajectory in the modified model is almost the same as before, so the two trajectories overlap. On the other hand, the H^0-proton particle is pushed slightly inward by the new z-bumps. It is about 0.6 cm closer to the axis at z = 100 cm where the particles suffered the most losses. The new z-bumps act like a small steering dipole, which has an integrated bending field of about 0.04 T m. This yields a bending angle of about 7 mrad at 1 GeV. In addition, we can see that the H^0-proton particle trajectory is very close to the old vacuum chamber, as indicated by the red dashed curve. The new vacuum chamber in that region is depicted by the green dashed curve, which is 2.95 cm away from the old one. This new boundary should provide good space for the H^0-proton particles and reduce their losses in the magnet.

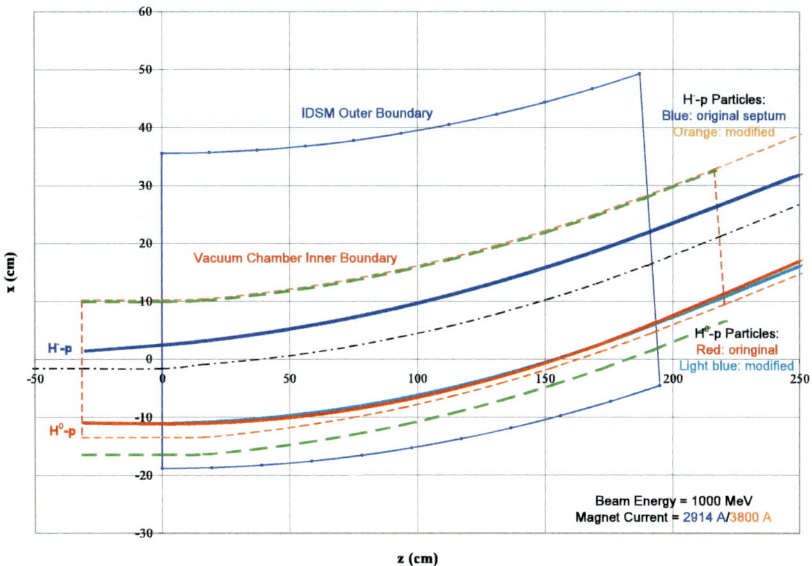

Figure 8.24. Waste beam trajectories in IDSM before and after modification.

The integrated bending fields of the modified septum are measured by a flip coil and compared with three-dimensional simulation results. The flip coil is a BNL product with a cross section of 1.27 x 1.27 cm^2 and a length of 365.76 cm. The coil has 20 turns of conducting wires in a single layer with a width of 1.2974 cm [25]. The measurement is performed along four selected tracks on the IDSM mid-plane. Each track follows the natural curvature inside the septum and is perpendicular to the magnet face outside. The four tracks are 10.4978, 5.4178, -3.7262, and -9.5682 cm from the IDSM axis, respectively. The measured integrated fields as a function of the magnet current are shown in Figure 8.25. The

H^0-proton particle trajectories go through the region between tracks 1 and 2. Its integrated field at 3800 A is about 1 T m. The H^--proton particle trajectories go through the region between tracks 3 and 4, with a maximum integrated field close to 0.8 T m at 3800 A. In the plot we also show the three-dimensional simulation data at 3800 A, as indicated by larger diamonds. The difference between the measurement and simulation is less than 1%. The agreement is quite good.

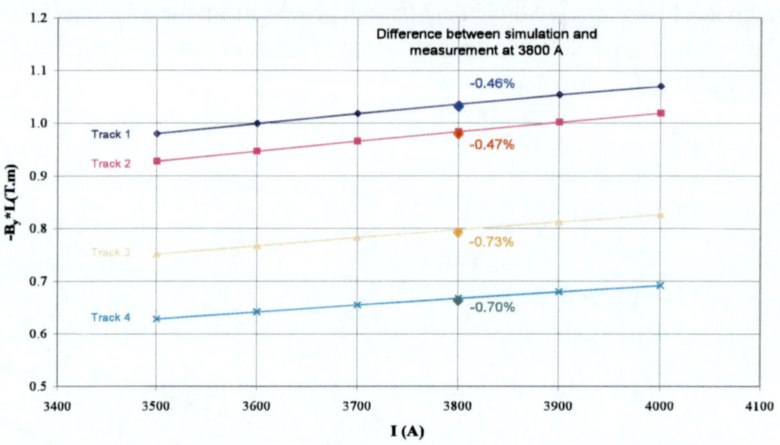

Figure 8.25. Integrated fields of the modified septum for 1 GeV operation.

As seen in Figure 8.2, the circulating proton beam line (light blue line marked by a letter "P") is very close to the IDSM. At the magnet entrance, the IDSM septum plate almost touches the vacuum chamber of the circulating proton beam. Magnetic fringe fields of the IDSM affect the proton beam and this problem should be taken care of. In the simulation of the old septum, the magnetic fringe field component B_y on the circulating proton beam axis is more than 6 G, while B_x has a peak of more than 14 G, as shown in Figure 8.26. This is higher than the required 1 G within the proton beam pipe. The integrated fringe field B_y along the proton beam axis is 1345 G cm, while the B_x integral is 110 G cm. Since the circulating proton beam occupies a rather large area, the IDSM fringe fields off the proton beam axis are even a bigger concern. Figure 8.27 shows the IDSM fringe fields within a circular area of radius R = 8 cm around the proton beam axis, where the axial position is at the septum entrance. The IDSM fringe fields are as high as 100 G at the locations closest to the septum. This is detrimental to the proton beam. In order to minimize the fringe effect, we have wrapped the proton beam vacuum chamber by two layers of CO-NETIC AA μ-metal [26]. The μ-metal shielding reduces the fringe fields inside the beam pipe to bellow 1 G.

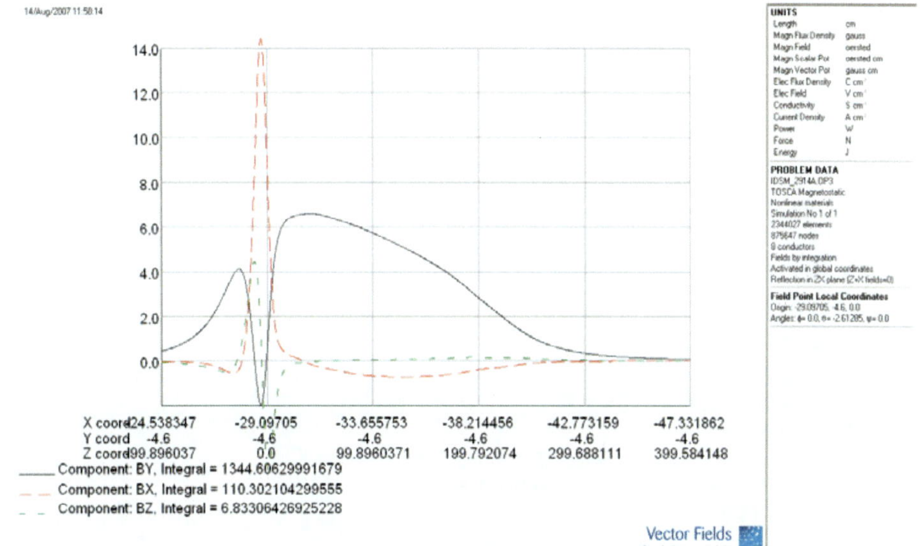

Figure 8.26. Magnetic fringe fields on the circulating proton beam axis in the old septum.

Figure 8.27. IDSM fringe fields within a circular area of radius R=8 cm around the proton beam axis.

For the modified IDSM, the situation would be much worse in terms of magnetic fringe field effect due to the enlarged gap and the new z-bumps. In the first step of the modification, the septum plate is not extended out. The fringe fields on the proton beam axis are measured by a Hall probe and compared with simulations, as shown in Figure 8.28. The agreement is good. However, the fringe fields are so high that it is not possible to reduce them to an acceptable level inside the vacuum chamber by two layers of μ-metal shielding. Thus, we extend the septum plate out by 25 cm. The fringe fields on the proton beam axis from simulation are also shown in Figure 8.28 by the red curve for B_y and the blue curve for B_x. With two layers of CO-NETIC AA μ-metal wrapped around the proton beam vacuum chamber, the fringe fields on the beam axis are much reduced to an acceptable level, as indicated by the red dots and blue diamonds from measurements.

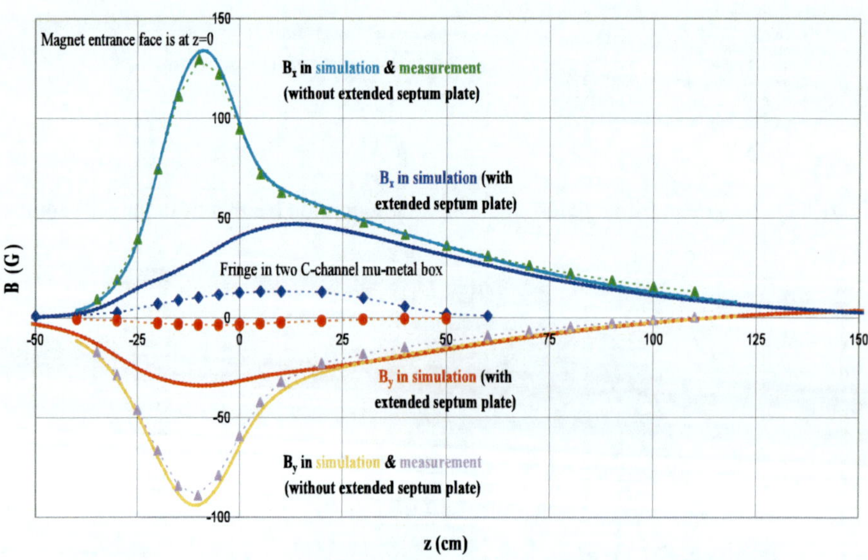

Figure 8.28. Magnetic fringe fields on the proton beam axis in the modified septum.

8.6.4. Waste Beam Losses Downstream after Quadrupole Magnet

The waste beam losses after the quadrupole magnet (30Q58) in the SNS ring injection waste beam dump line are caused by two factors: the equivalent phase space area occupied by the H^0-proton and H^--proton beams at the quad entrance is very large; and the focusing downstream is not adequate. One possible solution for the problem is to finely tune the two waste beams in order to largely reduce their equivalent phase space area at the entrance of the quad. This requires more knobs upstream for fine tuning. On the other hand, the easiest and most effective way to solve the transport problem after the quadrupole magnet (30Q58) is to add another quadrupole magnet such as a 30Q44 to form a quadrupole doublet. The doublet can reduce the beta function amplitude downstream and make the beam transport there possible. This approach is rather straightforward, and we skip the details.

8.7. Experimental Verifications

Beam measurements have been made to compare with model predictions. Here we report the results before and after the D4 move. The comparison between early measurements and simulations is qualitative, while recent measurements are more quantitative.

Before we produced preliminary results in three-dimensional simulation studies, many beam measurements during the commissioning and early operation were already performed. Measurements of the waste beam parameters indicated qualitative agreement with the simulation models. For example, since the SNS ring commissioning the particle losses along the entire injection dump line were very high, especially at the IDSM. Activation measurements showed elevated levels about half-way along the length of the IDSM on the beam right side (looking downstream), indicating that the H^0-proton waste beam was in fact scraping on the IDSM vacuum chamber. This was in good agreement with the trajectories shown in Figure 8.9.

Another example was the vertical motion of the H^--proton beam in experiments. Beam position measurements showed a 42 to 47 mm vertical separation of the H^0-proton beam and the H^--proton beam at a Beam Position Monitor BPM01 located immediately downstream after the quadrupole magnet. The simulation models with three different chicane dipole settings, i.e., the production, delivered, and design settings, yielded a vertical separation of the two waste beam centroids at the same location in the range of 30 to 73 mm. Both the measurements and the simulations resulted in the same order of magnitude for the H^--proton vertical motion at BPM01. No further simulation effort was made to better compare the BPM01 signal at the exact experimental conditions, such as beam energy, injection parameters, etc, in addition to the chicane dipole settings.

More quantitative comparison between measurements and simulations has been made after chicane dipole D4 was moved by Δx = 8 cm. New diagnostics such as a Wire Scanner WS00 and a BPM00 have also been installed just after the IDSM. The coordinates of WS00 and BPM00 in the models are (78.15, 2.3, 834.11) and (106.53, 2.3, 961.44). Figure 8.29 shows the measurement from WS00, where the two waste beams are separated by 149 mm in the horizontal plane.

A new simulation model with the same chicane setting, same beam energy, and same injection conditions has been built. This model yields a horizontal separation of 136 mm for the two waste beam centroids at the WS00 location. The discrepancy is about 9%. Considering all the errors in the experimental setup and the diagnostic parameters, the agreement between the WS00 measurement and the new simulation result is quite good. The new model also predicts a vertical separation of 2.4 mm for the two waste beam centroids at the WS00 location. This is too small to be accurately measured by WS00. As an independent verification of the vertical motion, measurements at BPM00 show an H^0-proton beam and H^--proton beam vertical separation of 4.5 mm. The new simulation model yields 3.2 mm for this separation at the BPM00 location, again in good agreement with the measurement considering the 1 - 2 mm measurement error. In comparison with the situation before the D4 move, the vertical motion of the H^--proton beam has been dramatically reduced by one order of magnitude.

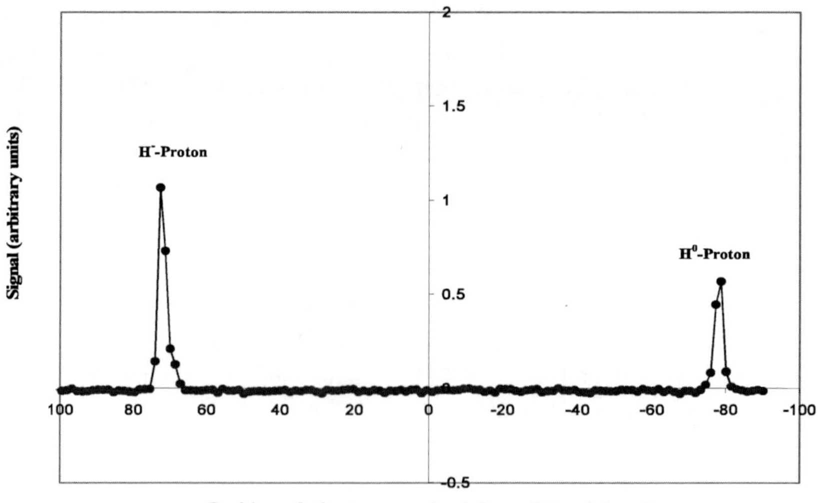

Figure 8.29. Horizontal separation of two waste beams measured by WS00.

With all the remedies implemented, the waste beam particle losses in the IDSM are reduced to an acceptable level. Indeed, this problem no longer exits for the SNS ring injection operation.

REFERENCES

[1] R. L. Martin, “History of non-Liouvillean injection and its connection to the initiation of the U.S. program in heavy ion fusion,” in AIP Conference Proceedings No. **253**, p. 232, New York, 1991.

[2] Hofmann, “Non-Liouvillean method applied to heavy ion fusion,” ibid, p. 149.

[3] J. Wei, D. T. Abell, J. Beebe-Wang, M. Blaskiewicz, P. R. Cameron, N. Catalan-Lasheras, G. Danby, A. V. Fedotov, C. Gardner, J. Jackson, Y. Y. Lee, H. Ludewig, N. Malitsky, W. Meng, Y. Papaphilippou, D. Raparia, N. Tsoupas, W. T. Weng, R. L. Witkover, and S. Y. Zhang, “Low loss design for the high-intensity accumulator ring of the Spallation Neutron Source,” *Phys. Rev. ST Accel. Beams* 3, 080101 (2000). (http://prst-ab.aps.org/abstract/PRSTAB/v3/i8/e080101)

[4] J. Wei, J. Beebe-Wang, M. Blaskiewicz, J. Brodowski, A. Fedotov, C. Gardner, Y.Y. Lee, D. Raparia, V. Danilov, J. Holmes, C. Prior, G. Rees, S. Machida, “Injection choice for Spallation Neutron Source ring,” in Proceedings of the 2001 Particle Accelerator Conference, Chicago, IL, June 18–22, 2001, eds. P. Lucas and S. Webber (IEEE Piscataway, NJ, 2001), p. 2560.

[5] D. Raparia for the SNS collaboration, “SNS injection and extraction devices,” in Proceedings of the 2005 Particle Accelerator Conference, Knoxville, TN, May 16–20, 2005, ed. C. Horak (IEEE Catalog Number 05CH37623C, 2005), p. 553.

[6] J. Wei, “Preliminary change request for the SNS 1.3 GeV-compatible ring,” in BNL/SNS Tech Note 76, May 4, 2000.

[7] D.T. Abell, Y.Y. Lee, and W. Meng, “Injection into the SNS accumulator ring: minimizing uncontrolled losses and dumping stripped electrons,” in Proceedings of the Seventh European Particle Accelerator Conference, Vienna, Austria, June 26–30, 2000, p. 2107.

[8] D. Raparia, A. Jain, W. Meng, Y.Y. Lee, and J. Jackson, “Magnet analysis and engineering support,” in ASAC (Accelerator System Advisory Committee) Review of the SNS Project, September 2004.

[9] M. A. Plum, "Injection dump review: history and the path forward," in SNS injection dump beam line review, Nov. 21, 2006.

[10] J. Error of SNS/ORNL, private communication.

[11] J. G. Wang, "3D modeling of particle losses in SNS ring injection dump beam line," in Proceedings of the 18th Meeting of the International Collaboration on Advanced Neutron Sources, Dongguan, China, April 25–29, 2007; also see J. G. Wang, SNS-NOTE-MAG-173, February 28, 2007.

[12] J. G. Wang, "3D modeling of SNS ring injection dump beam line," in Proceedings of the 2007 Particle Accelerator Conference, Albuquerque, NM, June 25–29, 2007 (IEEE Catalog Number: 07CH37866, 2007), p. 3660; also see J. G. Wang, SNS-NOTE-MAG-178, August 21, 2007.

[13] J. G. Wang and M. Plum, "Three-dimensional particle trajectories and waste beam losses in injection dump beam line of Spallation Neutron Source accumulation ring," *Phys. Rev. ST Accel. Beams* 11, 014002 (2008). (http://prst-ab.aps.org/abstract/PRSTAB/v11/i1/e014002).

[14] J. Simkin of Vector Fields, England, private communication.

[15] J. G. Wang, "Magnetic field distribution of injection chicane dipoles in Spallation Neutron Source accumulator ring," *Phys. Rev. ST Accel. Beams* 9, 012401 (2006). (http://prst-ab.aps.org/abstract/PRSTAB/v9/i1/e012401); also see J. G. Wang, SNS-NOTE-MAG-130, 2004.

[16] N. Tsoupas, "SNS magnets: modeling and measurements," in ASAC (Accelerator System Advisory Committee) Review of the SNS Project, February 2002.

[17] J. G. Wang, "Field distributions and particle trajectories in SNS ring injection dump septum magnet," SNS-NOTE-MAG-170, September 7, 2006.

[18] M. Reiser, *Theory and Design of Charged Particle Beams,* John Wiley and Sons, Inc., New York, 1994, Chapter 5.

[19] OPERA-3D User Guide and Reference Manual, Vector Fields Software of Cobham Technical Services, Oxford, England; also see http://www.vectorfields.com/

[20] N. Tsoupas, J. Brodowski, W. Meng, J. Wei, Y. Y. Lee, and J. Tuozzolo, "A large-aperture narrow quadrupole for the SNS accumulator ring," in Proceedings of the Eighth European Particle Accelerator Conference, Paris, France, June 3–7, 2002, eds. T. Garvey et al. (EPS-IGA and CERN, 2002), p. 1106.

[21] N. Tsoupas, J. Jackson, Y. Y. Lee, D. Raparia, and J. Wei, "Magnetic field calculations for a large aperture narrow quadrupole", in Proceedings of the

2003 Particle Accelerator Conference, Portland, OR, May 12–16, 2003, eds. J. Chew, P. Lucas, and S. Webber (IEEE Piscataway, NJ, 2003), p. 2153.

[22] D. Raparia, Y. Y. Lee, J. Wei, and S. Henderson, "Beam dump optics for the Spallation Neutron Source," ibid, p.3416.

[23] J. Holmes of SNS/ORNL, private communication.

[24] J. G. Wang, "Modification of a spare septum magnet for SNS ring injection dump beam line," in Proceedings of the Eleventh European Particle Accelerator Conference, Genoa, Italy, June 23–27, 2008, eds. I. Andrian et al. (EPS – AG, 2008), p. 2389; also see J. G. Wang, SNS-NOTE-MAG-181, March 28, 2008.

[25] J. Jackson of BNL, private communication.

[26] Magnetic shield corporation, see http://www.magnetic-shield.com/

Chapter 9

SNS Ring Extraction Lambertson Septum Magnet

The SNS ring Extraction Lambertson Septum (ELS) magnet contained a strong skew quadrupole term, which was identified as the source of causing a beam profile distortion on the target. We have performed three-dimensional computer simulations to study the magnetic field quality in the magnet. The skew quad term is computed with different methods in simulations and is compared to measurement data. The origin of the large skew quad term is thoroughly investigated and the remedy for minimizing the integrated skew quad term by modifying the magnet is proposed. Particle tracking is performed to show intuitively the beam profile evolution through the original and modified septum. The transfer matrix elements of the septum are computed to verify the beam profile. The magnetic interference to the septum performance from an adjacent quadrupole is also assessed.

9.1. Beam Profile Distortion Due to Strong Skew Quad Term in ELS

A Lambertson septum magnet is a device to be used to inject or extract the beam from above or below the plane of particle accelerators. In the SNS accumulator ring, the ELS is employed to extract the beam from below the plane of the ring. The SNS accumulator ring extraction system is built in one of its four straight sections. Figure 9.1 shows a part of the extraction straight section in top view. It contains an ELS magnet followed by a quadrupole doublet assembly.

During the ring accumulation, a circulating proton beam passes through a well-shielded aperture within the upper yoke of the ELS. Its effect on the beam should be negligible. The ring extraction takes place in a single turn and two steps after the beam is fully accumulated. Four pulsed kickers upstream (not shown in figure) are first fired. This pushes the beam down to the dipole entrance of the ELS, which in turn bends the beam horizontally out of the ring.

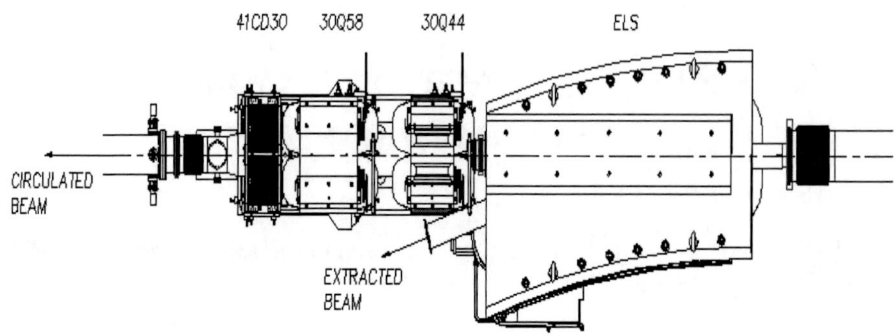

Figure 9.1. ELS and quad assembly (Courtesy of BNL design drawing and specifications).

The ELS was designed and developed at BNL [1 - 3]. Its structure is quite complex and its field qualities are very critical to the extracted beam as well as to the circulating beam. Although the ELS extraction line was originally measured at BNL with a flip coil on grids before delivery, there were no official test data of harmonic contents provided.

Since the SNS ring commissioning we have found that the extracted beam profile on the target is slightly tilted, as shown in Figure 9.2 [4]. This tilted beam profile does not match the target geometry and has been an issue of concern for high power operation. Considerable amount of effort has been devoted to finding the source of the beam profile tilt. Finally, three-dimensional simulation studies of the ELS field qualities show a strong skew quadrupole term in the ELS extraction line, and this has been identified to be responsible for the beam profile tilt on the target. Since the magnet is already installed for operation, there is no way to conduct further measurements on the magnet. This has motivated us to carefully model the magnet in three-dimensional computer simulations in order to accurately calculate the skew quad term and to improve its field quality by minimizing the skew quad term.

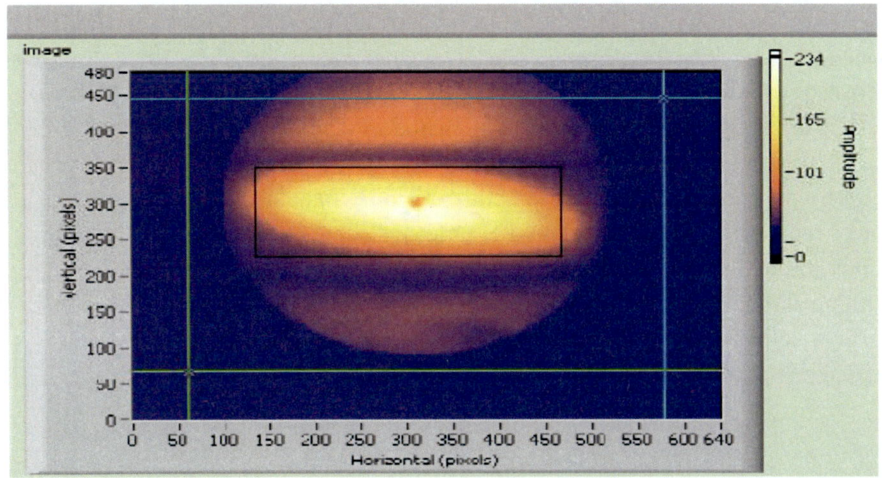

Figure 9.2. Beam image from the target view screen [4], where the target size is 20 cm by 7 cm.

In this chapter we first introduce our simulation model of the ELS in Section 9.2. Its general performance characteristics for the circulating beam and extracted beam are reported in Section 9.3. The computation of the skew quad term with three different methods and a comparison with the BNL measurement data are detailed in Section 9.4. In Section 9.5 we investigate the origin of the large skew quad component. This is caused by the lack of mid-plane symmetry in the magnet. The contributions to the skew quad term from the ELS central region and the two ends are analyzed. In Section 9.6 the remedies for minimizing the integrated skew quad term are proposed. Section 9.7 provides a beam tracking study, which shows clearly a beam profile tilt in the original ELS and its elimination in the modified one. In Section 9.8 we compute the transfer matrix elements of the septum by the method of test particles. The results can verify the beam profiles in Section 9.7. The magnetic interference to the septum performance from an adjacent quadrupole is assessed in Section 9.9.

9.2. Three-Dimensional Simulation Model

The simulation environment in this work is OPERA-3D/TOSCA v. 12. The version 12 can operate in computers with 64 bits and allows the users to employ unlimited elements in models. The ELS model is shown in Figure 9.3(a). A cross section at the magnet center and in perpendicular to the extracted beam axis is shown in Figure 9.3(b). The model is built with the OPERA package "Modeller" instead of the "Pre-processor". This makes it easier to study the interference between two magnets. In the model, mechanical dimensions follow the BNL design drawings and specifications, as listed in Table 9.1. The circulating beam passes through the upper, round, straight aperture, which is very well shielded by the septum plate and the end structures. The extracted beam goes through the lower, curved gap between the ELS pole-tips. On the pole-tip two longitudinal shims are added to improve the bending field uniformity. The ELS is installed with a roll angle of 2.55° and a pitch angle of 0.53° in order to accommodate the vertical velocity of the extracted beam at the magnet entrance. These features are neglected in the simulation, but this should not affect the field quality study.

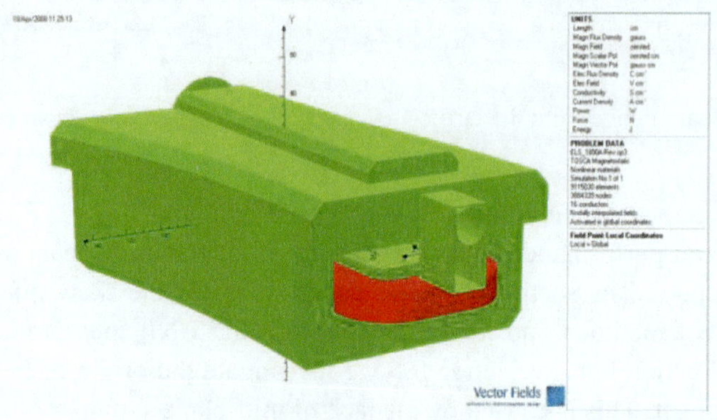

Figure 9.3. (a) Entrance view of ELS simulation model.

Table 9.1. ELS main design parameters [1 - 3]

Pole length (m)	2.23063
Pole width (m)	0.48006
Septum plate length (m)	2.63042
Septum plate width (m)	1.44196
Dipole gap for extracted beam (m)	0.16988
Effective (hard edge) length (m)	2.44
Bending angle (degree)	16.8
Bending field @ 1 GeV (T)	0.68
Aperture for circulating beam (m)	0.175

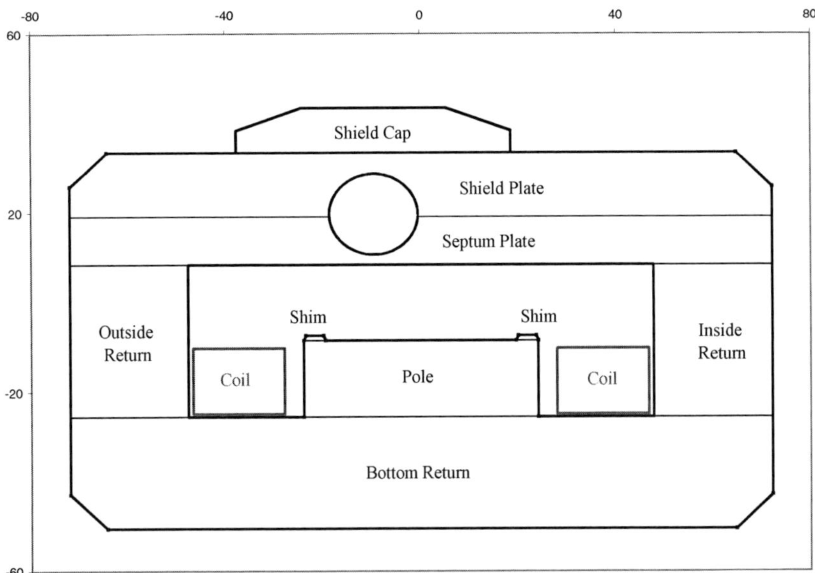

Figure 9.3. (b) Upstream view of a cross section through the center of ELS model.

In the model, the coordinate system origin is such that z = 0 is at 12.116 cm away from the pole-tip end on the magnet entrance side, y = 0 is at the mid-plane of the bending dipole, and x = 0 is at the circulating beam axis. The magnet is energized at 1890 A in the model, which is a nominal design value for a 1 GeV proton beam. The model consists of fine meshes for accurate results. It contains approximately ten million elements and its post-processor file occupies more than 2 GB memory. The B-H curve used in the simulations is a default one in OPERA-3D code. The septum is constructed with low carbon steel AISI 1006, but its exact B-H curve is not available. We believe that the error thus introduced for the main field should be less than a quarter of a percent according to our previous experience with other SNS magnet simulations and measurements.

9.3. General Performance of ELS

The SNS ring extraction septum combines two general functions in one magnet. First, it provides a straight path inside the upper yoke for the circulating beam. The main criterion for this function is to make the magnetic field along the path a minimum. The beam aperture is 17.5 cm and the minimum septum thickness is 1 cm. The circulating beam pipe is made of 1010 steel, and about 1 mm copper shielding is placed between the pipe and the magnet iron. In order to reduce the effect of the dipole fringe fields below, the septum plate is made approximately 20 cm longer than the pole-tip at each end. In addition, a shielding box is attached to the septum entrance face, and a round cap of iron is attached to the septum exit face. Second, the magnet bends the beam horizontally out of the ring during the extraction process. The main bending parameters, such as the bending angle, the nominal bending field, and the effective (hard edge) length, are listed in Table 9.1. Good field quality is essential for the performance of the extracted beam. In particular, the skew quad term should be small enough in order not to distort the extracted beam profile.

Figure 9.4 shows a zone map of the total magnetic flux density B_{mod} across the circulating beam aperture at the magnet longitudinal center (z = 123.428 cm). The field amplitude at different locations is indicated by the color index bar on the left side. Within the beam aperture, the magnetic field is at 0.5 G level. The field amplitude in the surrounding yoke is as high as more than 1.2 T. This demonstrates the very effective shield of the magnetic beam pipe with a surrounding gap of 1.181 mm, in addition to the reduction of residual fields in the circulating beam aperture of a Lambertson septum structure.

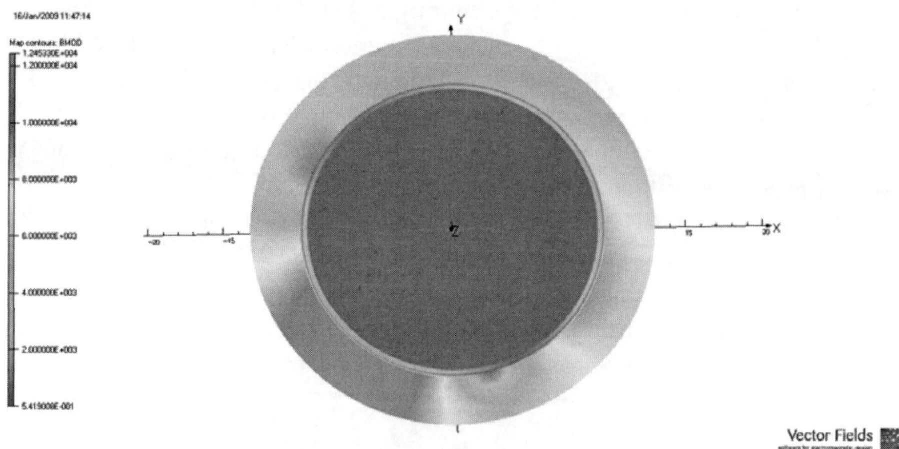

Figure 9.4. Magnetic fields across circulating beam aperture at magnet center.

The magnetic field distribution along the circulating beam axis in the ELS is shown in Figure 9.5. Although a shielding box at the septum entrance and a round cap of iron at its exit are applied, the fringe fields on the circulating beam axis in these regions still do not vanish. The integrated B_y component is about -0.0027 T m, while the integrated B_x is 0.0012 T m. This results in a total integrated dipole field of 0.0029 T m, which can be compensated by dipole correctors downstream.

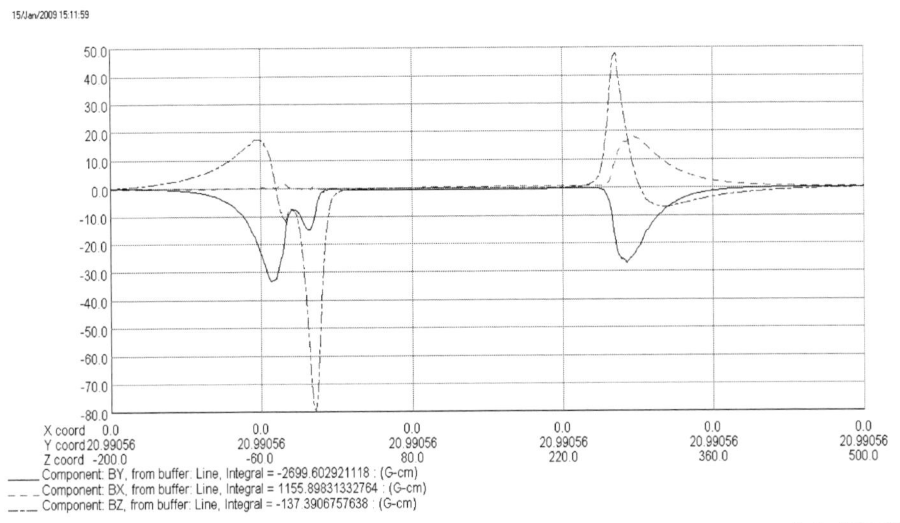

Figure 9.5. Magnetic field distribution on circulating beam axis.

By employing a rotation patch, which has the dimensions of r_1 = - 4.191, r_2 = 8.262, and z = -200 to 500 cm as explained in Section 2.6, we can evaluate the field within the circulating beam aperture. The method yields an integrated dipole field of 0.0029 T m with a phase angle of -157°. This is in fact the same value as that obtained on the circulating beam axis, as shown in Figure 9.5. The integrated quad term is 0.006 T with a phase angle of 65.9°. The magnetic field parameters of the ELS along the circulating beam line were also measured at BNL with a rotating coil. Their data show that the integrated dipole component at 1890 A is 0.0032 T m, which is slightly larger than our simulated value.

For the extracted beam, the magnetic field quality is also studied in the post-processor file. Figure 9.6 shows the magnetic fields across the ELS gap at its longitudinal center, where the local coordinate origin is x = 9.064, y = 0, z = 123.428 cm, and the three Euler's angles are 0, 8.4°, and 0, respectively. The dipole field B_y is quite uniform within the beam aperture region with a magnitude of 6687.9 G at its center. Figure 9.7 shows the magnetic bending field along a reference trajectory for the extracted beam. Inside the hard edge, the reference track follows the designed, curved ELS axis, while it is straight outside the hard edge. The total integrated dipole field along the reference trajectory is 1.692 T m.

This results in a hard edge length of 2.53 m and a bending angle of 17.1° for 1 GeV proton beams, which is slightly larger than the nominal value of 16.8°.

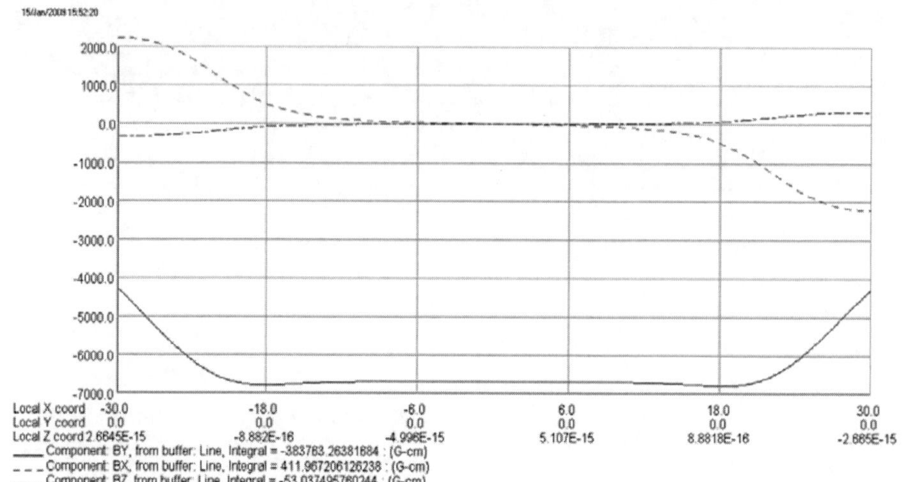

Figure 9.6. Magnetic fields across ELS gap at its center.

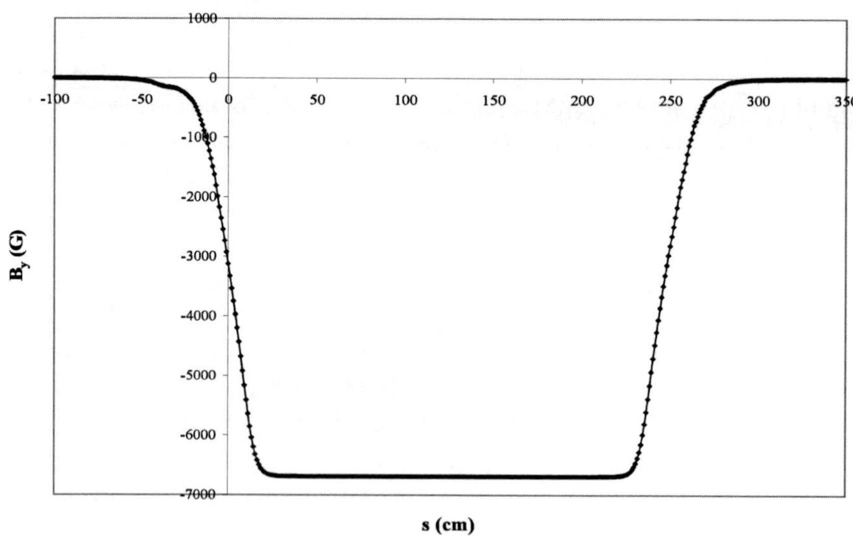

Figure 9.7. Magnetic dipole field on reference trajectory for extracted beam.

9.4. Computation of Skew Quad Term

The skew quad term in the ELS extraction channel is computed with three different methods: the patch rotation method, the surface field analysis, and the dipole field on grids. Different methods are employed in computation of the same parameter because of cross-checking purposes for gaining confidence in the modeling. In addition, the simulation results can be compared directly with the BNL measurement data.

9.4.1. Patch Rotation Method

The patch rotation method, as described in Section 2.6, is used to calculate the magnetic flux through a rotating Cartesian patch in a magnet. The rectangular patch in this calculation has a transverse dimension of r_1 = - 4.191 to r_2 = 8.362 cm. Since the reference trajectory inside the hard edge region is curved, we have to use piece-wise straight patches. There are a total of six patches employed. The first patch is for the upstream region with a length of z = -150 to 0 cm and its axis coincides with the z-axis. The sixth one is for the downstream region with a length of 200 cm and starts at x = 36.062, y = 0, and z = 244.208 cm. Its axis forms an angle of 16.8^{o} with respect to the z-axis. Inside the hard edge region, the four patches uniformly divide the whole arc length. Each of the four patches has a length of 61.922 cm and their Euler's angles are phi = psi = 0, and theta = 2.1, 6.3, 10.5, and 14.7 degrees, respectively. In this way, we approximate each 4.2^{o} arc inside the hard edge region by a straight segment. We have tried more than four patches inside the hard edge region, but found no significant improvement in the result.

The total magnetic flux through the six patches as a function of the rotation angle is shown in Figure 9.8. It looks roughly like a sinusoidal signal for a dipole. The peak flux amplitude, which occurs at θ = 0 and 180°, is about 0.21 T m^{2}. A Fourier decomposition of this signal yields all the harmonics, where a reference radius R = 8 cm is used. Table 9.2 shows the first five of them. The table also includes the data from the other methods described in more detail later. The integrated normal dipole term is B_1*L = -1.6925 T m, which is in the minus y-direction and bends 1 GeV protons by 17.14^{o}, which is slightly higher than the designed value of 16.8^{o}. These numbers are very close to those derived from Figure 9.7. The integrated total quadrupole term accounts for 131 units. For a transport line magnet, the conventionally required field quality is that the higher harmonics should be less than 10 units. Thus, the total quadrupole term in the septum is more than one order of magnitude higher than the conventional requirement. The integrated skew quad term is A_2*L = -0.0214 T m. Dividing the value by the radius of R = 8 cm yields an integrated gradient of 0.267 T.

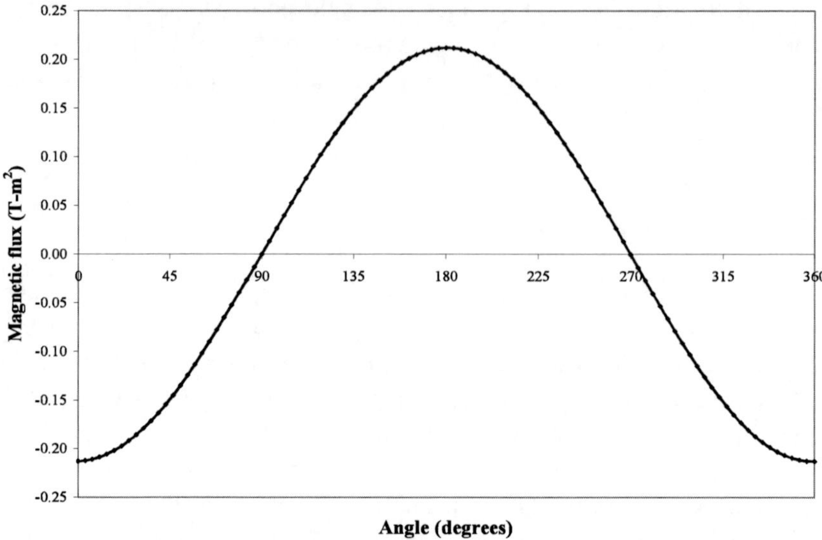

Figure 9.8. Total magnetic flux through the patches vs. rotation angle.

Table 9.2 Integrated ELS harmonics in T m from various methods

	Patch rotation		Surface field		Dipole field on grid		BNL test	
m	B_m*L	A_m*L	B_m*L	A_m*L	B_m*L	A_m*L	B_m*L	A_m*L
1	-1.6925	-0.0022	-1.6926	-0.0017	-1.6916		-1.7055	
2	-0.0059	-0.0214	-0.0066	-0.0215	0.0008	-0.0221	-0.0005	-0.0166
3	0.0027	-0.0007	0.0027	-0.0001	0.0023	0.0001	0.0077	0.0011
4	0.0001	-0.0050	0.0002	-0.0050	0.0002	-0.0051	-0.0012	-0.0022
5	0.0001	-0.0001	0.0001	-0.0001	0.0002	0.0001	-0.0025	-0.0001

9.4.2. Surface Field Analysis

This method is employed to calculate the magnetic field component B_r or B_θ on a cylindrical surface in order to obtain the z-dependent harmonic distribution of magnetic fields along a reference trajectory. The surface field analysis is in fact the method of three-dimensional multipole expansion described in Chapter 4.

The key procedure in this method is first to draw a circle, which is centered at and perpendicular to a prescribed reference trajectory. The OPERA-3D command CIRCLE is then employed to compute the radial component $B_x{*}t_y{-}B_y{*}t_x$ or the azimuthal component $B_x{*}t_x{+}B_y{*}t_y$ of the magnetic fields on the circle. The data are fit to the Fourier series for harmonic contents. A command input file is developed to move the circle in desired steps along the entire length of the reference trajectory and to accomplish the calculation automatically. The generalized gradients can be obtained through the three-dimensional multipole expansion. Note that t_x and t_y are the tangential unit vectors to lines in the OPERA-3D. Although B_r and B_t are also used in the OPERA-3D post-processor for the radial and azimuthal components of magnetic fields, they represent different physical quantities if the reference trajectory is curved.

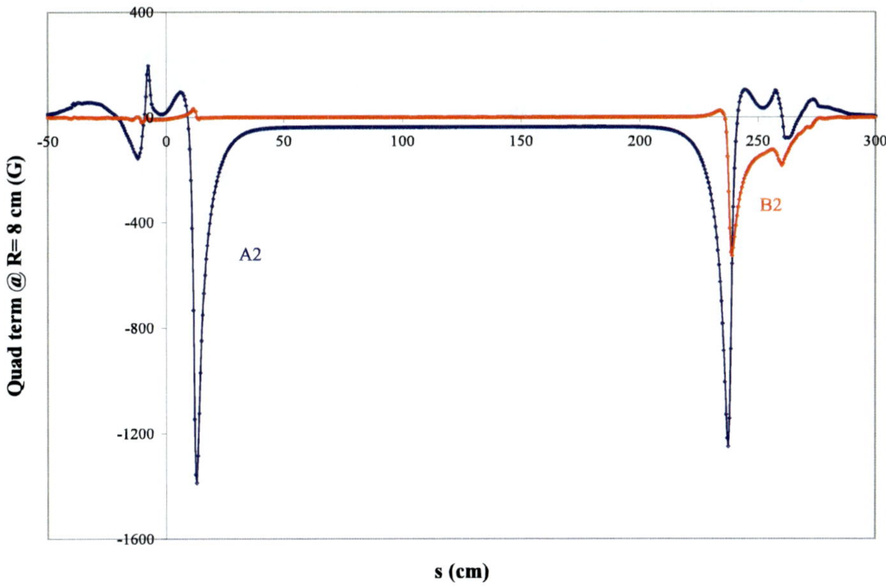

Figure 9.9. Quad term of magnetic fields on a cylindrical surface surrounding a reference track.

Figure 9.9 shows the quadrupole term of magnetic fields on a cylindrical surface of radius R = 8 cm along the entire reference track. The reference track here consists of a straight segment for the upstream region from z = -150 to z = 0 cm, an arc segment for the hard edge region from z = 0 to z = 244.208 cm, and a straight segment for the downstream region with a length of 200 cm. The integrated skew quad term is A_2*L = -0.0215 T m, resulting in an integrated gradient of 0.269 T. In comparison with the result from the patch rotation method, the difference is less than 0.5%. The integrated dipole term from this method is B_1*L = -1.6926 T m, yielding a bending angle of 17.14^o for a 1 GeV proton. The integrated skew quad term accounts for 127 units, which is fairly close to that from the patch rotation method. More harmonic data from this method are also listed in Table 9.2.

At the magnet center, where the local coordinate system origin is (9.064, 0, 123.428) and the Euler's angles are (0, 8.4, 0), the magnetic fields on a circle perpendicular to the reference track contain a dipole term C_1 = 6687.92 G and a skew quad term A_2 = -36.90 G, which accounts for 55.2 units. The skew quad term everywhere in the central region of the magnet is very close to this value. This is too high for a transport line magnet with only two-dimensional (x and y) field distributions.

9.4.3. Dipole Fields on Grids

This method calculates the bending fields along different, parallel trajectories and fits the results to a harmonic notation. The method mimics a flip coil measurement in experiments. The trajectories form a grid on a transverse cross section of the magnet aperture, as shown in Figure 9.10. There are a total of 21 positions in the grid for the integrated B_y calculation. The trajectory for each grid position consists of three parts: two straight lines before and after the hard edge region and one arc inside the hard edge magnet. All the trajectories are in parallel with the reference trajectory. The integrated bending fields at different grid points from the simulation as well as from the BNL measurement are listed in Table 9.3. The measurement data are normalized to a septum current of 1890 A, which is used in the simulation.

Table 9.3 Integrated bending fields on the grids

Point	r (cm)	Theta (^{o})	B_y*L (T m)	
			Simulation	BNL test
0	0.00		-1.6916	-1.7055
1	0.10	0	-1.6908	-1.7061
2	0.20	0	-1.6891	-1.7035

Point	r (cm)	Theta (°)	B_y*L (T m)	
3	0.30	180	-1.6915	-1.7034
4	0.40	180	-1.6907	-1.7001
5	0.50	-90	-1.7006	-1.7101
6	0.60	-46.32	-1.7007	-1.7087
7	0.70	-27.64	-1.7022	-1.7091
8	0.80	226.32	-1.7014	-1.7100
9	0.90	207.64	-1.7037	-1.7109
10	1.00	-90	-1.7084	-1.7235
11	1.10	-64.48	-1.7095	-1.7224
12	1.20	244.48	-1.7096	-1.7247
13	1.30	90	-1.6831	-1.6979
14	1.40	46.32	-1.6817	-1.6959
15	1.50	27.64	-1.6775	-1.6938
16	1.60	133.68	-1.6823	-1.6958
17	1.70	152.36	-1.6785	-1.6928
18	1.80	90	-1.6775	-1.6990
19	1.90	64.48	-1.6754	-1.6971
20	2.00	115.52	-1.6758	-1.6964

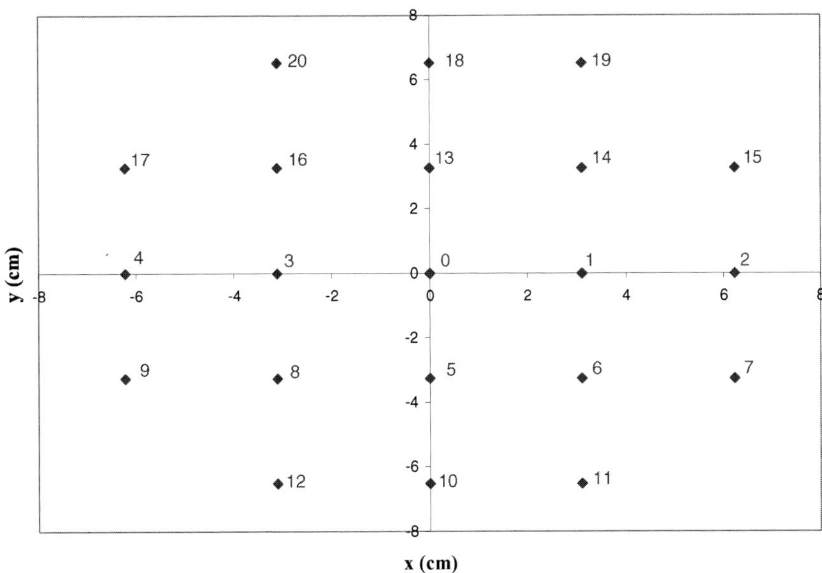

Figure 9.10. Grid points for the method of dipole fields on grids.

In cylindrical coordinates, a two-dimensional bending field B_y is given by Eq. (2.15b) and is rewritten here as

$$B_y(r,\theta)=\sum_{m=1}^{\infty} C_m\left(\frac{r}{R}\right)^{m-1} Cos[(m-1)\theta-m\alpha_m]. \tag{9.1}$$

A five term presentation of the field is

$$B_y(r,\theta)-C_1\cdot Cos(\alpha_1)=C_2\left(\frac{r}{R}\right)Cos(\theta-2\alpha_2)+C_3\left(\frac{r}{R}\right)^2 Cos(2\theta-3\alpha_3)+$$
$$+C_4\left(\frac{r}{R}\right)^3 Cos(3\theta-4\alpha_4)+C_5\left(\frac{r}{R}\right)^4 Cos(4\theta-5\alpha_5)+... \tag{9.2}$$

These equations also apply to the integrated bending fields, in which the field amplitude terms have the dimension of T m rather than T. It can be seen that the term $C_1{*}L\cdot Cos(\alpha_1)$ can be obtained by measuring the bending field on the reference trajectory (r = 0). Figure 9.11 plots the left hand side values of Eq. (9.2), $\Delta B_y{*}L = B_y(r,\theta){*}L - C_1{*}L\cdot Cos(\alpha_1)$, at different grid points based on the simulation calculation, as listed in Table 9.3. The figure also shows the BNL test data from Table 9.3, which will be discussed more in the next subsection. A best fitting of the simulation data by the least square method in MATHEMATICA [5] leads to the integrated field amplitude $C_m{*}L$ and phase angle α_m. The integrated normal and skew terms can then be found from Eqs. (2.14a, b). This method yields an integrated skew quad term $A_2{*}L$ = -0.0221 T m from the simulation data. This amplitude is larger by about 3% than that obtained with the patch rotation method and the surface field analysis. The other harmonics from the fitting are also listed in Table 9.2.

It should be noted that this method is not as accurate as the other two above. The main problem is that the fitting in MATHEMATICA is based on the least square method rather than the orthogonal condition. The fitting results do depend on the number of terms used in Eq. (9.1), especially for higher harmonics. Our practice shows that the quadrupole term remains roughly the same when two to six terms are used for the fitting. Although it is not as accurate and reliable as the patch rotation method and the surface field analysis, the method of the dipole fields on grids is developed here mainly for processing the BNL test data as described below.

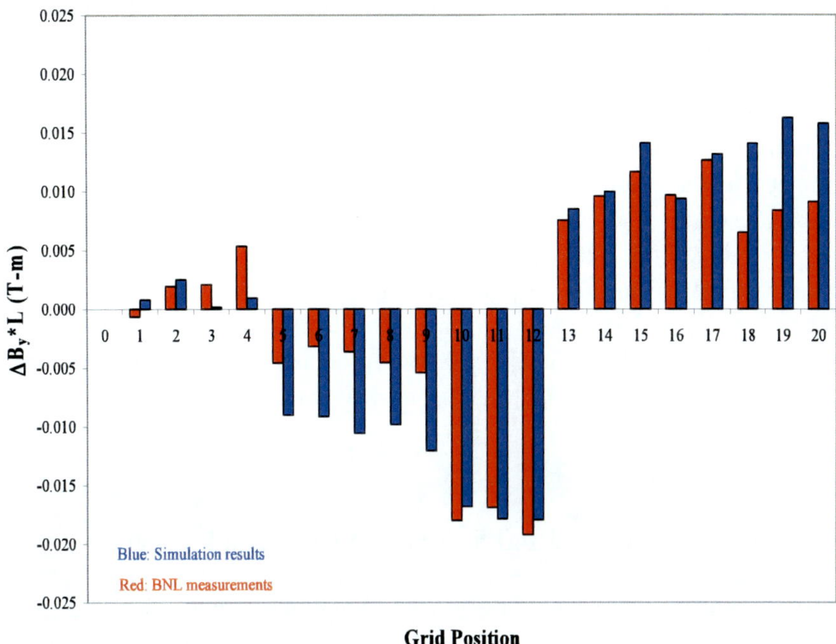

Figure 9.11. Simulation and test data from the integrated dipole field on grids.

9.4.4. Comparison with Measurements

The ELS extraction line was measured at BNL with a flip coil on the same grid pattern as in Figure 9.10. The BNL test data are listed in Table 9.3 and plotted in Figure 9.11. The same fitting method yields an integrated skew quadrupole term A_2*L = -0.0166 T m. The other harmonics from fitting the measurement data are listed in Table 9.2.

The integrated skew quad term in the BNL measurement data is only 75% of the value from the three-dimensional simulation. This large discrepancy between the three-dimensional simulation and the BNL measurement is likely explained by measurement errors. In measurements with a flip coil following a curved reference trajectory, it is very difficult to achieve a relative accuracy of 0.25% [6], which corresponds to an error of more than one vertical division in Figure 9.11. Although this discrepancy exists, the measurement data indeed show an integrated skew quad term of about 100 units, which is still one order of magnitude larger than that usually required for a transfer line magnet.

9.5 ORIGIN OF THE LARGE SKEW QUAD TERM

The integrated skew quad term in the ELS has two constituents as shown in Figure 9.9. The first one comes from the central region (s ~ 50 to 200 cm) of the septum magnet, which accounts for 25.8% of the total. This is due to the up-down asymmetry of the pole tip and the coil configuration. The original design did not optimize the compensation by the longitudinal shims. The second one results from the end effect of the magnet, which uses a longer septum plate to shield the circulating beam from the fringe fields of the dipole below. This structure naturally creates a field gradient in the vertical direction and contributes to the skew quad term, which accounts for 38.8% of the total at the magnet entrance and 35.4% of the total at the magnet exit. Thus, we investigate the two constituents separately.

9.5.1. Effect of Longitudinal Shims on Skew Quad Term

We use OPERA-2D to study the effect of longitudinal shims on the skew quad term in the central region of the magnet. A half cross section of Figure 9.3(b) is built in a two-dimensional model. The circulating beam aperture is left out in the model and this should not affect the simulation accuracy since the septum plate is thick enough. The simplified two-dimensional simulation model can speed up the process of finding an optimized shim geometry for a minimum or desired skew quad term.

Figure 9.12 shows a zoom-in view of the ELS pole region in the two-dimensional model, where the red block is a part of the conducting coil. The magnetic flux lines indicate the field uniformity. Due to the up-down asymmetry of the pole-tip and the coil configuration, the magnetic flux lines originated from the pole-tip tend to spread horizontally when they go up to the septum plate. They are denser just above the pole-tip than in the upper region of the gap. This creates a field gradient in the vertical direction, which is a skew quad term. It is a common practice to employ the shims for compensation of this field non-uniformity in the vertical direction in order to minimize the skew quad term. Unfortunately, the original design of the shim, which has a trapezoidal cross section with a bottom width of 4.496 cm, a top length of 3.683 cm, and a height of 1.125 cm, does not optimize the compensation and leaves a large skew quad term in the central region of the magnet.

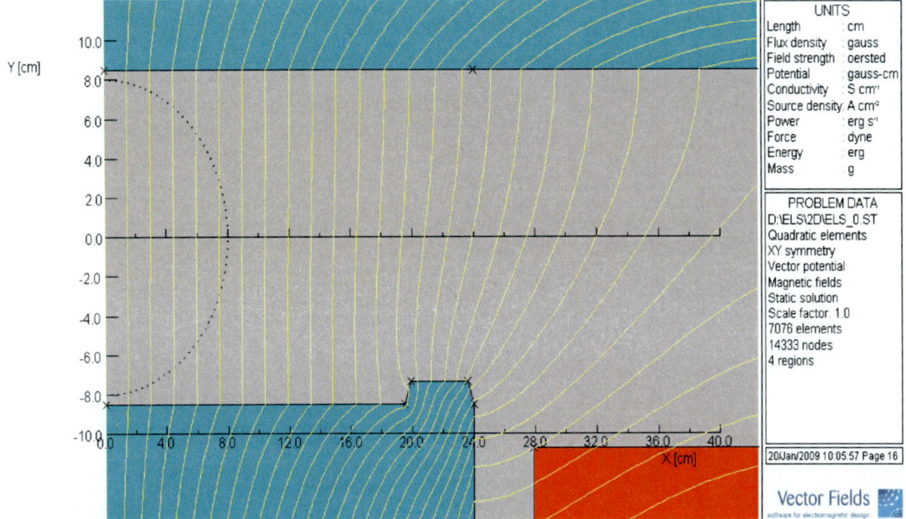

Figure 9.12. The central area of a two-dimensional ELS model.

In Figure 9.13 we plot the magnetic field component B_y of the original design along a vertical line from y = -8 to y = 8 cm at x = 0 (blue curve). It is clear that there is a gradient in the y-direction and the maximum difference in B_y is about 50.8 G. On a semicircle of radius R = 8 cm and centered at x = y = 0, as shown by the dots in Figure 9.12, we calculate the radial component $B_r = B_x*n_{vx}+B_y*n_{vy}$ or the azimuthal component $B_\theta = B_x*t_{vx}+B_y*t_{vy}$ of the magnetic fields. The data are decomposed into the Fourier harmonics, which are listed in Table 9.4. The total dipole field is -6692.30 G, while the skew quad term accounts for -37.79 G, corresponding to 56.5 units. This is fairly close to the result of 55.2 units for the central region of the magnet as obtained in the three-dimensional model as shown in Figure 9.9. Note that n_{vx} and n_{vy} are the normal unit vector to a line, and t_{vx} and t_{vy} are the tangential unit vector to a line in OPERA-2D.

Table 9.4 Harmonic amplitudes (in G) in three different shim cross sections

	Original design		Minimum A2		Over-compensated	
m	B_m	A_m	B_m	A_m	B_m	A_m
1	-6692.30	0.00	-6693.82	0.00	-6693.32	0.00
2	0.00	-37.79	0.00	-2.02E-03	0.00	+61.97
3	-2.51	0.00	-10.77	0.00	-14.44	0.00
4	0.00	-13.94	0.00	-1.51	0.00	21.36
5	-1.86	0.00	-7.69	0.00	-10.44	0.00

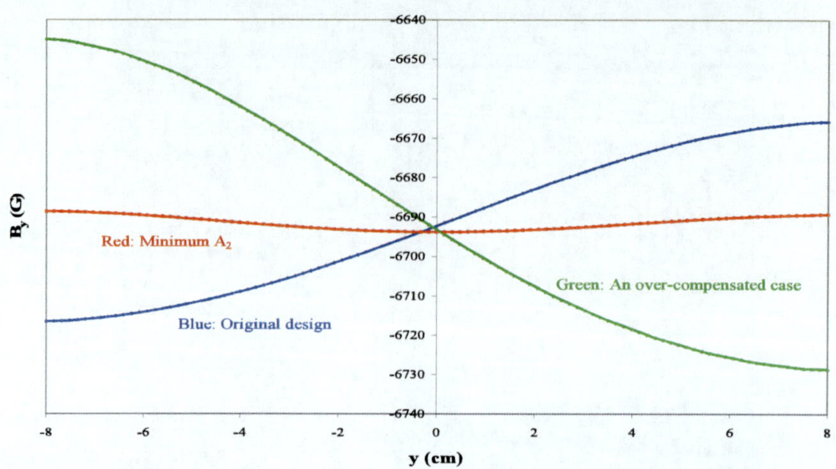

Figure 9.13. B_y vs. y at x = 0 for y = -8 to +8 cm in three different shim geometries.

The skew quad term in the ELS central region is very sensitive to the shim geometry. With the two-dimensional code we have generated various skew quad terms for different shim cross sections. Note that the specific shim cross section for a desired skew quad term is not unique. One may use the traditional method of trial and error to find a solution, or may employ the OPERA-2D OPTIMIZER to speed up calculations. In Figure 9.14 we plot three shim cross sections, including the original design. The field uniformity B_y vs. y for the other two shim geometries are also shown in Figure 9.13, and the corresponding harmonic contents are listed in Table 9.4.

The red shim geometry has a trapezoidal cross section with its bottom width of 7.431 cm, height of 1.143 cm, and 18° on its two edges. The magnetic field B_y along a line from y = -8 to y = 8 cm at x = 0 as shown in Figure 9.13 is quite flat and is almost an even function with respect to y = 0, indicating that no skew quad term exists. Indeed, a Fourier analysis of the data on the semicircle yields a skew quad term of -0.002 G, a negligible amount. By refining the shim geometry carefully, we should be able to eliminate completely the skew quad term in the central region of the magnet.

The green shim geometry also has a trapezoidal cross section with its bottom width of 5.842 cm, height of 2.54 cm, and still 18° on the two side edges. It would produce a field distribution along the y-axis, as plotted in Figure 9.13. The field at the bottom of the pole tip (y = -8 cm) is lower than that on the top (y = 8 cm), while it is the opposite in the original design. The maximum difference in B_y is 83.7 G. The same Fourier decomposition of the field data on the semicircle yields a skew quad term of +61.97 G. Note that the sign changes from the original case. This is the consequence of over-compensation by the shim. The other higher harmonics are listed in Table 9.4. The usefulness of this shim geometry will be clear in Section 9.6.

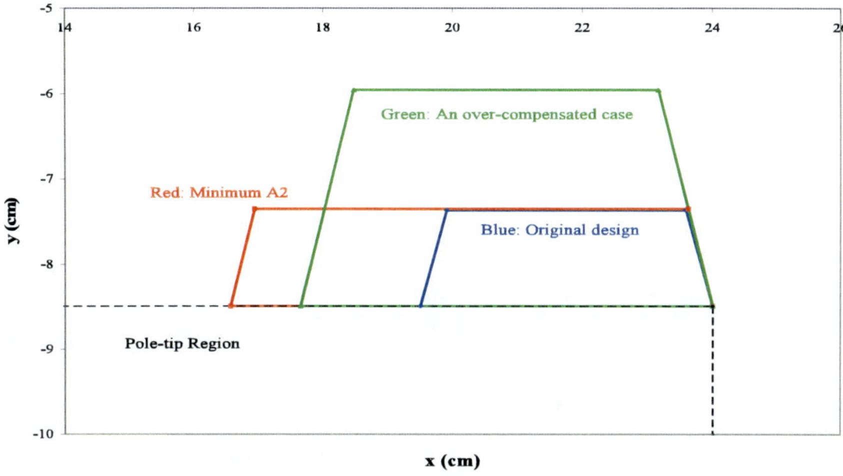

Figure 9.14. Cross sections of three longitudinal shims.

9.5.2. Effect of Magnet Ends on Skew Quad Term

In the ELS design, the septum plate is extended out of the pole-tip end by about 20 cm for shielding the circulating beam from the fringe fields of the dipole below. This structure naturally contributes to a skew quad term since there is an up-down field gradient in the fringe regions. The magnetic flux lines coming out of the pole-tip end spread out rapidly when they go up towards the septum plate. This is demonstrated in Figure 9.15, which shows the magnetic field component B_y versus y at x = 0 and z = 12.116 cm. The vertical line picked here for the plot is just at the septum dipole entrance, corresponding to the s-position for the first large peak in Figure 9.9. The field component B_y has an amplitude close to 1.1 T at y = -8 cm. The field amplitude decreases going up in the y-direction, and it drops to about 0.54 T at y = 8 cm. The field gradient is most profound in the lower half of the gap with a B_y difference of about 0.5 T from y = 0 to y = -8 cm, while the field is relatively flat in the upper half of the gap with a B_y difference less than 0.05 T from y = 8 to y = 0 cm. The B_y versus y plot at the same z = 12.116 cm and at the other x values from -8 to 8 cm is very similar to Figure 9.15.

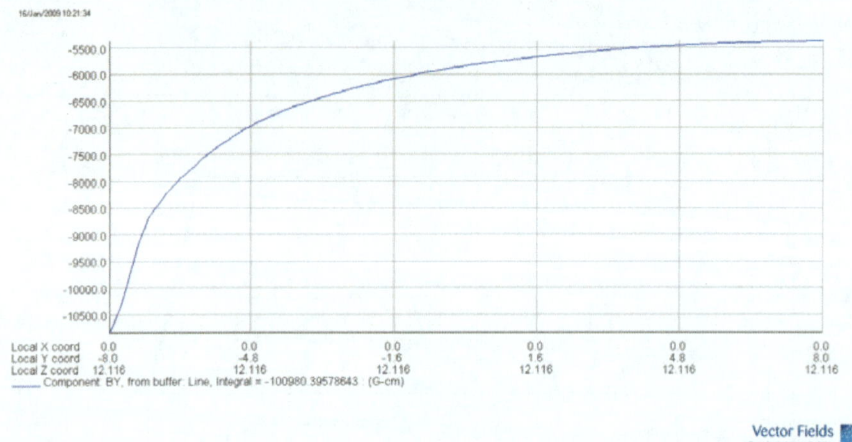

Figure 9.15. B_y vs. y at x = 0 and z = 12. 116 cm.

A chamfer angle at the pole-tip end would alleviate the problem, but it was overlooked in the original design. For the original ELS, we have noticed that there is a gap of more than 10 cm between the pole-tip end and the coil. This space can be utilized for the insertion of z-blocks with appropriate chamfer angles. In addition, the longitudinal shims in the magnet central region should be extended to the end of the z-block with specially trimmed edges. Such a z-block is illustrated in Figure 9.16. This method could reduce the vertical field gradient in the dipole entrance and exit, as well as the integrated skew quad term in the whole fringe regions.

In order to demonstrate the usefulness of the z-blocks with appropriate chamfer angles, we have built simplified ELS models, which are straight, shorter, and have no circulating beam aperture with associated end structures. This greatly speeds up the iteration process for obtaining desired z-block geometries in simulation. Figure 9.17 shows the skew quad term in the magnetic fields on a cylindrical surface of radius R = 8 cm in the magnet entrance fringe region for three different pole-tip and z-block configurations. In the figure, the pole-tip end is at about z = 12 cm and the septum plate end is at about z = -8 cm. The blue curve is for the original design where there is no z-block attached. Note that this curve is not exactly the same as the one in Figure 9.9 because it comes from the simplified straight model rather than the real model. An integration of A_2 from z = -100 to 60 cm yields -7300 G-cm. The contribution from the longitudinal shims in the region from z = 12 to z = 60 cm is about -1700 G-cm. Thus, the integrated skew quad term due to the fringe fields is about -5600 G-cm.

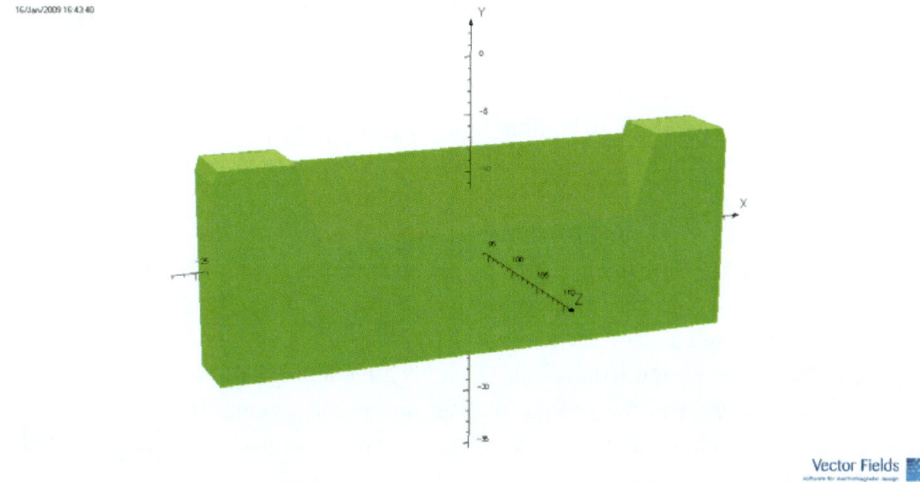

Figure 9.16. A z-block being attached to the pole-tip end for changing the integrated skew quad term.

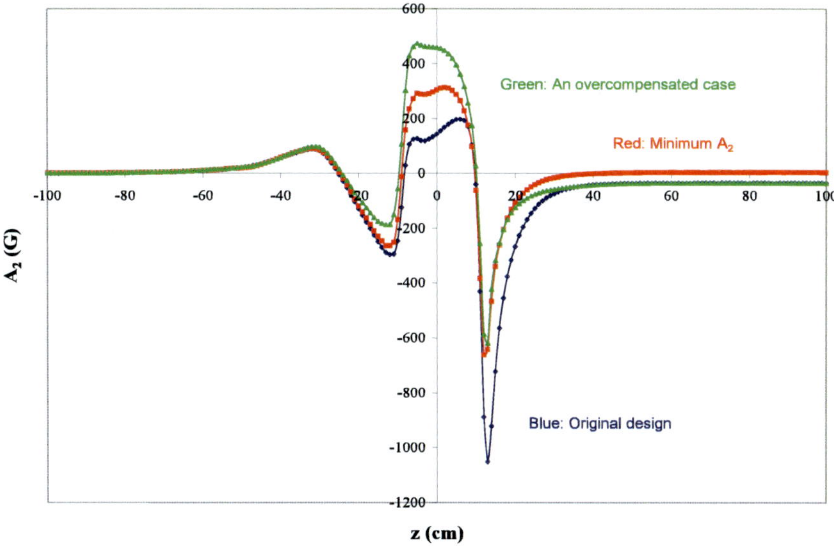

Figure 9.17. Skew quad term along a cylindrical surface of radius R = 8 cm for three cases.

The red curve shows the results from a modified simple model, where the longitudinal shims take the cross section as the red one in Figure 9.14 and a z-block is attached to the pole-tip end. The integrated skew quad term from z = -100 to 60 cm is now -5.5 G cm. Since the new shims do not contribute to the skew quad term inside the magnet, the -5.5 G cm results from the cancellation of the skew quad terms in the fringe region. This is a significant reduction for one end of the magnet. By carefully trimming the z-block, it is possible in principle to eliminate completely the integrated skew quad term.

The green curve shows the results from another simple model where the longitudinal shims remain as in the original design and the z-block takes a different geometry. The integrated skew quad term in the same region is +2737 G. This is an over-compensated case by the z-block that changes the sign of the integrated skew quad term. It turns out that this positive amount of the skew quad term in the fringe region would cancel the negative amount of the skew quad term in the central region of the septum magnet due to the original shims, resulting in an almost vanishing integrated skew quad term over the entire length of the simplified septum model.

9.6. Minimization of Integrated Skew Quad Term

The analysis results described in the previous section suggest that there are three possible options to modify the original ELS in situ in order to minimize its integrated skew quad term. The first option would be to replace the original longitudinal shims by the green one in Figure 9.14, which would produce a positive skew quad term in the central region to compensate the negative ones from the two ends. The second option would be to insert the z-blocks between the pole-tip end and the coil to produce a positive integrated skew quad term, which would cancel the contribution from the magnet central region. These z-blocks should be similar to the one that yields the green curve in Figure 9.17. The third option would be to employ the red shim geometry in Figure 9.14, which would make the skew quad term uniformly vanishing in the magnet central region. In addition, the z-blocks would be added to the pole-tip ends to produce a vanishing integrated skew quad term in the fringe region, similar to the red curve in Figure 9.17.

In principle, all these three options should lead to very small integrated skew quad term in a modified ELS. Nevertheless, the second and third options with the insertion of z-blocks are more involved in practice. The simplified ELS models with the z-blocks contain significantly higher integrated sextupole term. The insertion of the z-blocks, which is equivalent to an extension of the pole tip length, would increase the fringe field contribution to the circulating beam line and so more shielding is required. Furthermore, it is very time-consuming to optimize the z-block geometry in realistic three-dimensional ELS models, where the z-blocks at the magnet entrance and exit should be slightly different. We have so far produced a number of modified, realistic ELS models with the z-blocks, and their overall performance is not as good as that from the first option. Therefore, our proposed modification is the first option, which is also practically implementable in situ. This option is further investigated in more detail and is described below.

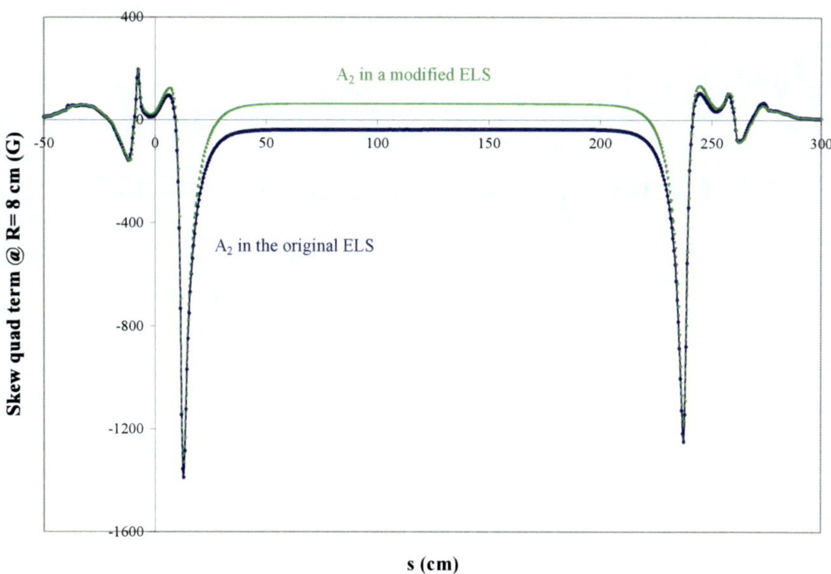

Figure 9.18. Skew quad term of the fields on a cylindrical surface surrounding a reference track.

A full three-dimensional modified ELS model based on the first option has been constructed. The new model requires only a change of the longitudinal shims, which take the green geometry in Figure 9.14. The method of the surface field analysis is employed to evaluate the harmonic contents of the new septum magnet. The skew quadrupole term of the magnetic fields on a cylindrical surface of radius 8 cm is shown in Figure 9.18. For comparison purpose, we also plot the result for the original ELS from Figure 9.9. The integrated skew quad term for the modified ELS is now 0.00054 T m, or 3.2 units. The corresponding integrated gradient is 0.0067 T. This is about 40 times smaller than that in the original magnet. It is expected that the integrated skew quad term would be even smaller when an adjacent quadrupole downstream of the ELS is taken into account, as discussed in Section 9.9.

The integrated higher harmonics for the original ELS are listed in Table 9.2. They are 16.1, 29.3, and 0.4 units for the sextupole, octupole, and decapole, respectively. For the modified ELS from the first option, the integrals of the sextupole term and octupole term are 1.4 and 19.6 units, respectively. Both of them are smaller than those in the original design due to the cancellation between the central region and the fringe regions of the septum magnet. The integrated decapole term in the modified septum is 10.3 units, larger than before. We do not expect any problem for the extracted beam from this slightly higher decapole term in the modified septum magnet.

9.7. Extracted Beam Profile through ELS

A skew quadrupole has its focusing and defocusing planes rotated by 45° with respect to the focusing and defocusing planes of a normal quadrupole. The governing equations of motion in a skew quad are coupled second-order differential equations. This was already explained in Section 2.5. Thus, the skew quadrupole distorts the beam profile in a lattice conventionally consisting of all the normal quadrupoles.

In order to verify the effect of the skew quad term on the beam profile in the SNS ring extraction Lambertson septum magnet, we study the extracted beam trajectories and profiles through the ELS before and after the modification of the magnet. We launch a laminar beam of an elliptical cross section without space charge upstream of the magnet. The initial particles move in the direction of the z-axis. We select a reference particle on the axis and 36 more on the surface of the ellipse. Since the integrated ELS bending field in the models is slightly higher than the nominal design value, the proton energy is set to 1.0298 GeV in order for the reference particle to follow the nominal design trajectory. OPERA-3D TRACK command is employed to calculate the particle trajectories through the ELS extraction channel as shown in Figure 9.19. At the end of the trajectories, we create a Cartesian patch, which is perpendicular to the reference trajectory. The intersection of all the particle trajectories on the patch is recorded. This yields the extracted beam profile downstream of the ELS.

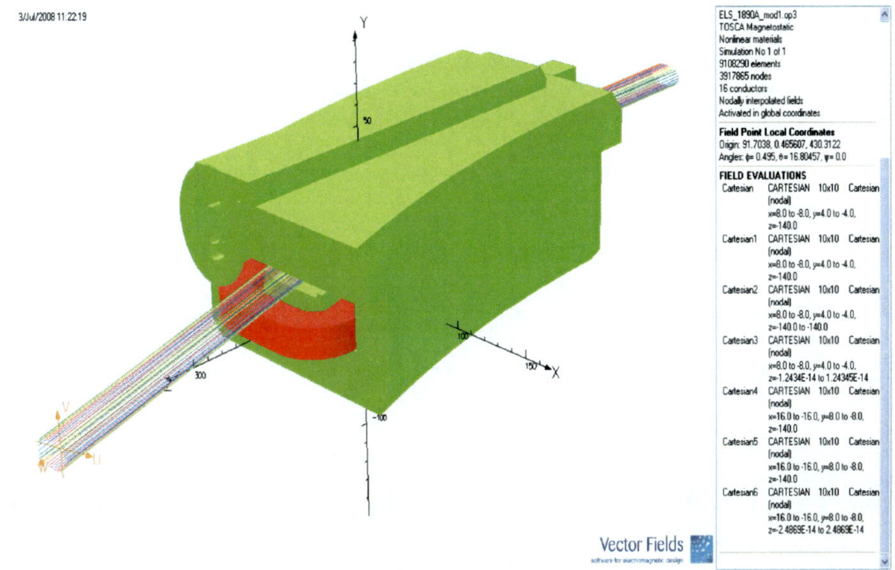

Figure 9.19. Trajectories of a laminar beam without space charge through the ELS.

Figure 9.20 shows the extracted beam profiles from the study. The dotted blue curve is the beam profile at z = -150 cm from the ELS entrance. It is a perfect, right ellipse, which should be the case in the nominal design and operation. The red curve indicates the extracted beam profile from the original septum at its downstream. The plot is made by the interception of the beam particles on a plane, which is perpendicular to the reference trajectory at a distance of 194.2 cm from the ELS exit. It is clear that the extracted beam profile is distorted and looks tilted. The test particle upstream at A(4, 0) on the elliptical surface is moved to A′ downstream of the original ELS. The angle between OA and OA′ is about 9.7°. The beam profile tilt appears to be even larger than this angle. This extracted beam profile resembles what observed in the early experiment as shown in Figure 9.2, though there are differences due to different beam parameters, location of evaluation, etc.

The green curve in Figure 9.20 shows the extracted beam profile from the modified septum at the same downstream location. The test particle at A is moved to the point A″. Very little beam profile tilt can be seen. Indeed, the extracted beam profile cross section is a fairly good ellipse, though its dimensions in both the x- and y-directions are slightly smaller than the initial ones. The profile tilt is corrected, or the x – y coupling is removed, with the ELS modification.

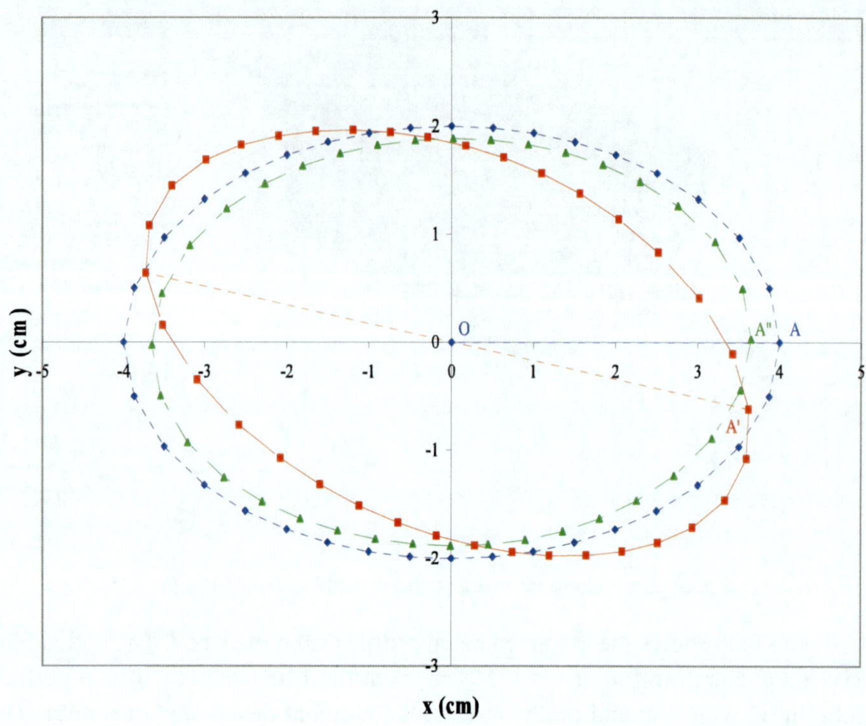

Figure 9.20. Extracted beam profiles through the original and modified ELS.

9.8. Linear Transfer Matrices

The ELS extraction channel contains not only the skew quad term, but also a normal quad term. This can be seen in Figure 9.9 where the red curve B_2 indicates that a small focusing quadrupole component appears at the septum entrance and a defocusing quadrupole component appears at the septum exit. These two focusing-defocusing components form a quadrupole doublet, which yields a net focusing effect on both x- and y-plane. That is why the green beam profile at the septum downstream in Figure 9.20 is smaller than the blue beam profile at the septum upstream. On the other hand, for the skew quad term in the septum, one often desires the coupling coefficients in lattice calculations in order to compute how much the beam profile is distorted. All of these can be done more analytically through the linear transfer matrices.

The transverse motion of particles in a general quadrupole can be described by the first-order **R**-matrix

$$\begin{bmatrix} x_2 \\ x'_2 \\ y_2 \\ y'_2 \end{bmatrix} = \begin{bmatrix} \mathbf{R}_{xx} & \mathbf{R}_{xy} \\ \mathbf{R}_{yx} & \mathbf{R}_{yy} \end{bmatrix} \begin{bmatrix} x_1 \\ x'_1 \\ y_1 \\ y'_1 \end{bmatrix} = \begin{bmatrix} R_{11} & R_{12} & R_{13} & R_{14} \\ R_{21} & R_{22} & R_{23} & R_{24} \\ R_{31} & R_{32} & R_{33} & R_{34} \\ R_{41} & R_{42} & R_{43} & R_{44} \end{bmatrix} \begin{bmatrix} x_1 \\ x'_1 \\ y_1 \\ y'_1 \end{bmatrix}. \quad (9.3)$$

By employing the method of test particles as described in Section 5.2, we can obtain all the sixteen matrix elements from particle trajectory data in simulation models. The minimum number of test particles is five, including one reference particle. The reference particle always starts on the beam axis.

The simplest way for calculation is to launch two test particles on the x-z plane and another two on the y-z plane. For the ELS calculation, the reference particle of 1.0298 GeV starts at x = y = 0 and z = -150 cm along the z-direction. A rectangular patch perpendicular to the reference trajectory is created downstream of the ELS. The patch in fact defines the transverse planes of test particles in a moving frame. The first test particle is launched at $x_1 = 2$ cm, $x_1' = y_1 = y_1' = 0$. Its intersection on the patch yields x_2, x_2', y_2, y_2', leading to the matrix coefficients R_{11}, R_{21}, R_{31}, R_{41}. The second test particle starts at $x_1 = 0$, $x_1' = \text{Tan } 2^o$, $y_1 = y_1' = 0$. The resulting x_2, x_2', y_2, y_2' lead to R_{12}, R_{22}, R_{32}, R_{42}. Similarly, two more test particles initiated on the y-z plane yield the matrix elements in the third and fourth column of the R-matrix.

The field calculation by integration is chosen for better accuracy in this practice. For the original design of the ELS, the numerical results of the **R**-matrix from this calculation is

$$\mathbf{R} = \begin{bmatrix} \mathbf{R}_{xx} & \mathbf{R}_{xy} \\ \mathbf{R}_{yx} & \mathbf{R}_{yy} \end{bmatrix} = \begin{bmatrix} 0.9537 & 5.8907 & -0.1491 & -0.3717 \\ -0.002234 & 1.0350 & -0.04714 & -0.1282 \\ -0.1468 & -0.3622 & 0.9436 & 5.6891 \\ -0.04484 & -0.1219 & -0.02987 & 0.8781 \end{bmatrix}. \quad (9.4)$$

The determinant of $\mathbf{R}_{xx}$ is 1.00021, while it is 0.9986 for $\mathbf{R}_{yy}$. Both the $\mathbf{R}_{xx}$ and $\mathbf{R}_{yy}$ matrices are for focusing. This is consistent with the observation from Figure 9.9 that a normal quadrupole doublet exists in the septum. As far as the two coupling matrices $\mathbf{R}_{xy}$ and $\mathbf{R}_{yx}$ concerned, their elements have rather significant numbers.

For the modified ELS, the **R**-matrix is

$$\mathbf{R} = \begin{bmatrix} \mathbf{R}_{xx} & \mathbf{R}_{xy} \\ \mathbf{R}_{yx} & \mathbf{R}_{yy} \end{bmatrix} = \begin{bmatrix} 0.9541 & 5.8873 & 0.006780 & 0.04003 \\ -0.002875 & 1.0306 & 0.001668 & 0.004288 \\ 0.007910 & 0.03774 & 0.9419 & 5.6863 \\ 0.003193 & 0.009499 & -0.03070 & 0.8766 \end{bmatrix}. \tag{9.5}$$

The determinant of $\mathbf{R}_{xx}$ is 1.00028, while it is 1.00025 for $\mathbf{R}_{yy}$. Both the $\mathbf{R}_{xx}$ and $\mathbf{R}_{yy}$ matrices are for focusing and this explains why the green profile in Figure 20 is smaller than the blue profile in both x and y direction. The focusing strengths in both models before and after modification are close. The elements in the coupling matrices $\mathbf{R}_{xy}$ and $\mathbf{R}_{yx}$ are quite different between (9.4) and (9.5). In the modified ELS, the coupling matrix elements are reduced by a factor of more than one order of magnitude. That is why the beam profile distortion is largely removed.

9.9. ELS Interference with Adjacent Quad

The magnetic interference in a quadrupole doublet assembly is presented in Chapters 6 and 7 and is also reported in [7]. For interference studies of a two magnet assembly in simulations, we need at least two models for comparison purpose. One is for an individual magnet and the other is for the two magnet assembly. Special caution should be taken to avoid artificial errors due to mesh generation in the two models. The best way to achieve this goal is that the same two magnet assembly model is used to simulate any individual magnet provided that the other magnet is presented by air material and zero-coil current [8].

The interference between the ELS and an adjacent quadrupole 30Q44 (see Figure 9.1) is investigated in a simulation model, consisting of the two magnets. As shown in Figure 9.21, the quadrupole 30Q44 center is about 300 cm downstream of the ELS entrance. The closest distance between the surface of the ELS exit cap and the 30Q44 yoke is merely 22.261 cm. It is expected that the quad 30Q44 affects the ELS extracted beam and vice versa. The simulation model contains about 15 million elements, and the post processor file occupies 4.2 GB memory.

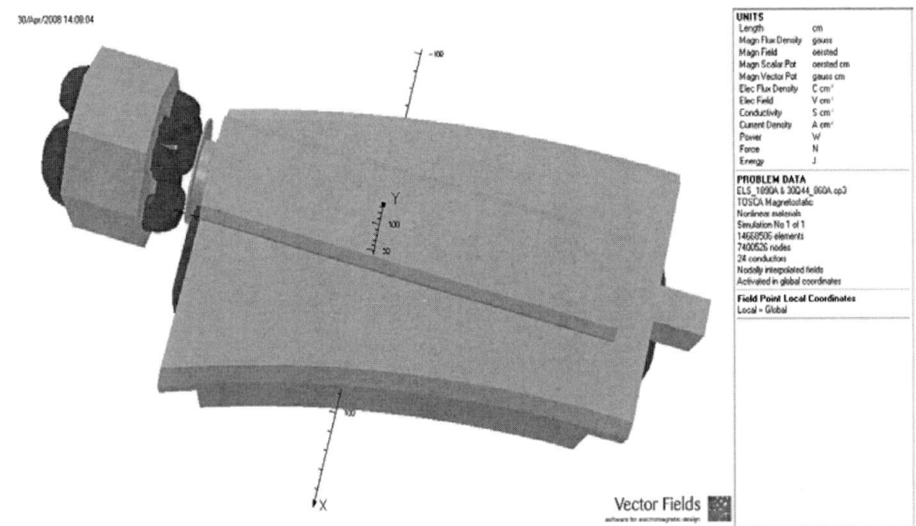

Figure 9.21. Simulation model of ELS and adjacent quad 30Q44.

9.9.1. Effect of the ELS on 30Q44

The existence of the ELS in the vicinity of 30Q44 changes the quadrupole performance parameters, as reported in [9]. By applying the method of patch rotation [Section 2.6] to the model in Figure 9.21, we can obtain two dimensional field parameters being compared with those for a single 30Q44, as listed in Table 9.5. With the presence of the ELS, the integrated gradient of 30Q44 is reduced by 1.3%. A dipole kick of 51 units (m = 1) is exerted on the quad, which accounts for about 840 G cm. In addition, other higher harmonics are generated for the quadrupole field, such as the sextupole of 9.62 units, etc.

9.9.2. Effects of 30Q44 on the ELS

The existence of 30Q44 in the vicinity of the ELS makes the magnetic flux lines originated from the ELS dipole region spread further to the quad yoke material. This is equivalent to increasing the vertical field gradient in the ELS exit region. The detailed analysis shows that the total integrated skew quad term in the original ELS alone is -0.0215 T m, while it is -0.0226 T m in the ELS and 30Q44 model. This is an increase of about 5%. As mentioned in Section 9.6, for the modified ELS the longitudinal shims with the green cross section shown in Figure

9.14 could reduce the total integrated skew quad term to +0.00054 T m. Here we intentionally make this value slightly positive. Since the existence of 30Q44 increases the vertical gradient of the field in the ELS exit region, the total integrated skew quad term goes in the negative direction in the modified ELS plus 30Q44 assembly. The numerical data show a total integrated skew quad term of A_2*L = -0.00033 T m, corresponding to an integrated gradient of G*L = 0.0042 T. This is a reduction of more than a factor of 60 in comparison with the original design.

Table 9.5 Interference in two dimensional parameters

	ELS+30Q44	30Q44 alone
Gradient integral (T)	2.0453	2.0713
	Harmonic Amplitude (units)	
m = 1	51.48	0.00
2	10000	10000
3	9.62	0.00
5	3.44	0.00
6	1.80	1.97
Reference radius (cm)	8	
Integral length (cm)	z = -100 to 100	
	* z = 0 is at 30Q44 center	

9.10. SUMMARY

The SNS ring Extraction Lambertson Septum magnet contained a strong skew quad term, which caused a beam profile distortion on the target. We have performed three-dimensional computer simulations with OPERA-3D/TOSCA to study the magnetic field quality in the magnet. The skew quad term is computed with various methods for better accuracy and reliability. The origin of the large skew quad term is investigated for the central region and the fringe regions of the magnet. We find that the skew quad term could be made uniformly vanishing in the central region by appropriate cross sections of the longitudinal shims; the integrated skew quad term in the fringe regions could also be made negligibly small with the z-blocks attached to the magnet pole-tip end. As a simpler alternative, a modification of the original ELS for reducing its integrated skew quad term is proposed, in which new longitudinal shims are employed to compensate the integrated skew quad term in the fringe regions. The modified ELS could reduce the integrated skew quad by a factor of 40, or more than 60 if the interference from an adjacent quad 30Q44 is taken into account. The study of the extracted beam trajectories and profile in the original and modified ELS demonstrates the benefits of such a correction. The computation of the linear transfer matrices by the method of test particles verifies what observed in the beam profiles. The detailed information on this work can be found in [10, 11].

The proposed remedy has been implemented in situ by replacing the two longitudinal shims with the new ones described in Section 9.6. Subsequent beam diagnostics show that the beam profile distortion indeed is gone and the problem has been solved [12]. It should be noted that the present modification is good only for the beam energy around 1 GeV. When the SNS accumulator ring goes to 1.3 GeV in the future power upgrade, the x-y coupling would reappear due to saturation of the longitudinal shims. The addition of appropriate z-blocks at the two ends of the modified ELS could be a solution for the future problem.

REFERENCES

[1] N. Tsoupas, Y. Y. Lee, J. Rank, and J. Tuozzolo, "Design of beam-extraction septum magnet for the SNS," in Proceedings of the 2001 Particle Accelerator Conference, Chicago, IL, June 18–22, 2001, eds. P. Lucas and S. Webber (IEEE Piscataway, NJ, 2001), p. 3245.

[2] J. Rank, K. Malm, G. Miglionico, D. Raparia, N. Tsoupas, J. Tuozzolo, and Y. Y. Lee, "Design considerations for a Lambertson septum magnet for the Spallation Neutron Source," in Proceedings of the 2003 Particle Accelerator Conference, Portland, OR, May 12–16, 2003, eds. J. Chew, P. Lucas, and S. Webber (IEEE Piscataway, NJ, 2003), p. 2150.

[3] D. Raparia, "SNS injection and extraction devices," in Proceedings of the 2005 Particle Accelerator Conference, Knoxville, TN, May 16–20, 2005, ed. C. Horak, (IEEE Catalog Number 05CH37623C, 2005), p. 553.

[4] M. A. Plum, "Commissioning of the Spallation Neutron Source accelerator systems," in Proceedings of the 2007 Particle Accelerator Conference, Albuquerque, NM, June 25–29, 2007 (IEEE Catalog Number: 07CH37866, 2007), p. 2603.

[5] Stephen Wolfram, The MATHEMATICA Book (Wolfram Media/ Cambridge University Press, 1999), 4th ed.

[6] A. K. Jain of BNL, private communication.

[7] J. G. Wang, "Particle optics of quadrupole doublet magnets in Spallation Neutron Source accumulator ring," *Phys. Rev. ST Accel. Beams* 9, 122401 (2006). (http://prst-ab.aps.org/abstract/PRSTAB/v9/i12/e122401).

[8] W. Z. Meng of BNL, private communication.

[9] J. G. Wang, "SNS ring extraction septum magnet and its interference with adjacent quadrupole," in Proceedings of the 2007 Particle Accelerator Conference, Albuquerque, NM, June 25–29, 2007 (IEEE Catalog Number: 07CH37866, 2007), p. 626.

[10] J. G. Wang, "Performance improvement of an extraction Lambertson septum magnet in the Spallation Neutron Source accumulator ring," *Phys. Rev. ST Accel. Beams* 12, 042402 (2009). (http://prst-ab.aps.org/abstract/PRSTAB/v12/i4/e042402)

[11] J. G. Wang, "3D field quality studies of SNS ring extraction Lambertson septum magnet," in Proceedings of the 2009 Particle Accelerator Conference, Vancouver, B.C., Canada, May 4–8, 2009, ed. M. Comyn, MO6PFP031.

[12] M. A. Plum, "SNS ring operational experience and power ramp up status," ibid, WE4PBC02.

INDEX

A

B

C

D

E

F

G

H

I

L

M

N

O

P

Q

R

S